W0091053

Compound Energy Systems
Optimal Operation Methods

RSC Energy Series

Series Editor:
Julian Hunt FRS, *University College London, London, UK*

Titles in the Series:
1: Hydrogen Energy: Challenges and Prospects
2: Fundamentals of Photovoltaic Modules and its Applications
3: Compound Energy Systems: Optimal Operation Methods

How to obtain future titles on publication:
A standing order plan is available for this series. A standing order will bring delivery of each new volume immediately on publication.

For further information please contact:
Book Sales Department, Royal Society of Chemistry,
Thomas Graham House, Science Park, Milton Road, Cambridge,
CB4 0WF, UK
Telephone: +44 (0)1223 420066, Fax: +44 (0)1223 420247, Email: books@rsc.org
Visit our website at http://www.rsc.org/Shop/Books/

Compound Energy Systems
Optimal Operation Methods

Shin'ya Obara
Department of Electrical and Electronic Engineering, Kitami Institute of Technology, Hokkaido, Japan

Arif Hepbasli
Mechanical Engineering Department, Ege University, Bornova, Izmir, Turkey

RSC Publishing

RSC Energy Series No. 3

ISBN: 978-1-84973-031-0
ISSN: 1757-6741

A catalogue record for this book is available from the British Library

Published by The Royal Society of Chemistry,
Thomas Graham House, Science Park, Milton Road,
Cambridge CB4 0WF, UK

Registered Charity Number 207890

For further information see our web site at www.rsc.org

Preface

Reduction of an environmental impact with energy consumption is an important issue in the world. However, since renewable energy is unstable, in many cases, it requires support by the conventional energy equipment using a fossil fuel. In order to increase the utilization rate of renewable energy, investigation of two methods is required. One is the development of highly efficient energy storage equipment represented by a battery and heat-storage tank. Another is development of the operation optimization technology of the compound energy system including green energy. It is thought that the energy supply method shifts from the individual operation of large-scale plant to distribution of small equipment. The change of such an energy supply method can introduce the optimal energy system into a region. As a result, it becomes possible to arrange the smallest energy system of an environmental impact. The way of thinking of this technology is connected with a microgrid or a smart grid. A microgrid and a smart grid require fusion of energy technology and an information technology. For example, the operation in consideration of the green energy change with load prediction and weather prediction of a compound energy system can be planned. Moreover, we should aim at realization of the Nature-grid that consists of green energy and information technology. This book describes the operation optimization technology by compound utilization of a fuel cell, photovoltaics, wind-power generation, hydrogenation engine, solar reformer, diesel power plant, *etc.* The technology described in this book plays a large role in the development of a small-scale power-generation system, a microgrid, and a smart grid, which are introduced into individual houses, apartment houses or local area power supplies. Further, the exergy analysis method, which is a very useful tool and has recently gained greater importance in the design, analysis, simulation and performance evaluation of various energy systems, is also included to deduce the potential for improvement.

Shin'ya Obara
Arif Hepbasli

RSC Energy Series No. 3
Compound Energy Systems: Optimal Operation Methods
By Shin'ya Obara and Arif Hepbasli

Published by the Royal Society of Chemistry, www.rsc.org

Contents

RSC Energy Series No. 3
Compound Energy Systems: Optimal Operation Methods
By Shin'ya Obara and Arif Hepbasli

Published by the Royal Society of Chemistry, www.rsc.org

CHAPTER 1

Background

SHIN'YA OBARA

1.1 Distributed Energy System

Distributed energy systems with sustainable energy operation have been widely discussed recently from the point of view reducing the environmental impact of society.[1–3] In these setups, the operation optimization program installed in the controller of a combined system is the most important aspect of the technology for determining the performance of the system.[4] However, because an output prediction for the green energy contribution to the system is required, the dynamic operation plan of a system that combines conventional energy equipment (for example, a diesel engine, a gas engine, a fuel cell, *etc.*) and green-energy equipment can be very difficult to design. In this work, we use a neural network (NN) to obtain output predictions for a solar cell. Weather data from the past 14 years (amount of solar radiation and outside temperature) is fed into the learning process of the NN. This NN production-of-electricity prediction algorithm (PAS) was developed by the author and is described in ref. 5. In this book, details of a compound energy system with the power prediction algorithm of green energy like the PAS are described.

Power fluctuations are known to occur in systems that utilize green energy on an independent microgrid and that experience large or rapid changes in load.[6] Given this, power storage equipment must be introduced and the dynamic characteristics of the microgrid must be improved. Due largely to the proliferation of hybrid vehicles and the like, the cost and performance of batteries have recently improved remarkably.[7] With this in mind, this book investigates algorithms for the operation planning of a microgrid that combines conventional energy equipment, a solar cell and a battery. Since a microgrid is

RSC Energy Series No. 3
Compound Energy Systems: Optimal Operation Methods
By Shin'ya Obara and Arif Hepbasli

Published by the Royal Society of Chemistry, www.rsc.org

typically built up of two or more energy systems, we have to solve a nonlinear problem with many variables. Therefore, this book shows the operation condition of generating equipment in chromosome code, and describes how to optimize operation for a compound energy system using a genetic algorithm (GA).

1.2 Independent Microgrid

The introduction to an urban area of a microgrid has the following advantages: (a) The heat transport distance is short and effective use of the exhaust heat of the generating equipment is possible; (b) The optimal facility for the energy demand characteristic of a community is installed, and a system having small environmental impact can be built; and (c) With an independent microgrid, the scale of equipment for distributing electricity is small.[8–10] Furthermore, (d) Connecting renewable energy considering regionality is expected to be an advanced system in microgrid technology. At present, the method of a microgrid interconnecting with commercial power, *etc.* is investigated (interconnect microgrid).[10] However, in order to achieve the advantages of (a) to (d) described above, it is necessary to operate a microgrid independently. The subjects of the independent microgrid are backup in the case of overload, and securing power quality (voltage and frequency). Furthermore, it is necessary to clarify the power-generation efficiency, the carbon-dioxide emissions, and the power cost of an independent microgrid. An improvement in power-generation efficiency is expected from the independent microgrid using a fuel cell compared with conventional electric power-supply technology. However, for the moment, fuel cells are expensive, and whether they will spread is not clear. As for a fuel-cell-independent microgrid, power-generation efficiency and carbon-dioxide emissions are expected to be advantageous compared with existing generating equipment. However, because the fuel cell is expensive, it is difficult to install the capacity corresponding to a load peak. Consequently, there is a case of operation that limits operation of a fuel cell to a highly efficient load region.[11] The hydrogenation technology of a city gas engine is effective concerning efficiency falls and increases in carbon-dioxide emissions at the time of partial load.[12–15] The power-generation system using a city gas engine with generator (NEG) is cheap compared with the fuel cell. Therefore, this book describes the investigation method of the power-generation efficiency and carbon-dioxide emissions in case of connecting NEG and PEFC (proton-exchange membrane fuel cell) to a microgrid.

1.3 Distribution Plan of Energy System

PEFC and SOFC (solid-oxide fuel cell) may develop as a power plant. These fuel cells have the advantage that they are highly efficient and have little environmental impact. However, these fuel cells are expensive, and the system is complex. It may be possible to reduce the number of expensive fuel cells that need to be installed by connecting the fuel cell to a microgrid and supplying power to two or more buildings. If the energy of the overall grid is supplied by one set of fuel cells (central system), the facility costs will be reduced considerably. Past work has examined the method of supplying power to a water electrolyzer and hydrogen and oxygen fuel storage methods.[16] Another study looked at controlling the number of units that divide a fuel cell and a reformer, finding that the system efficiency falls when operated at partial load.[17] In addition, energy-storage methods, such as batteries and flywheels, have been considered, though this equipment is not introduced in this book. Energy storage methods must consider power fluctuations in the microgrid affecting how the fuel cell is controlled. Although this is an important topic, for simplicity, this book focuses on other issues related to microgrids. There are no examples of the effect of the power demand pattern of buildings linked to a microgrid on the generation efficiency of a fuel cell system. This book will examine how the overall generation efficiency is affected by connecting a building to another grid. In this book, the fuel cell microgrid (FC microgrid) is installed in an urban area and divided into multiple grids. The system efficiency is improved over the case where each grid is connected independently (partition cooperation system). By dividing the grid and increasing the load factor of PEFC linked to each grid, the proposed method improves the generation efficiency of the overall grid. This book describes the generation efficiency of the FC microgrid using the power-demand model assuming typical buildings (individual houses, apartments, hotels, convenience stores, small offices, factory, and small hospitals).

References

1. Abu-Sharkh, Can Microgrids Make a Major Contribution to UK Energy Supply? *Renew. Sustain. Energy Rev.*, 2006, **10**(2), 78–127.
2. M. Muselli, G. Notton and A. Louche, Design of Hybrid-Photovoltaic Power Generator, With Optimization of Energy Management, *Sol. Energy*, 1999, **65**(3), pp. 143–157.
3. Y. Ismail, Y. Kemmoku, H. Takikawa and T. Sakakibara, An Operating Method for Fuel Savings in a Stand-Alone Wind/Diesel/Battery System, *J. Jpn. Sol. Energy Soc.*, 2002, **28**(2), 31–38.
4. S. Obara, Operating Schedule of a Combined Energy Network System with Fuel Cell, *Int. J. Energy Res.*, 2006, **30**(13), 1055–1073.
5. S. Obara and I. Tanno, Fuel Reduction Effect of the Solar Cell and Diesel Engine Hybrid System with a Prediction Algorithm of Solar Power Generation, *J. Power Energy Syst., JSME*, 2008, **2**(4), 1166–1177.
6. S. Obara, Dynamic Operation Plan of a Combined Fuel Cell Cogeneration, Solar Module, and Geo-Thermal Heat Pump System Using Genetic Algorithm, *Int. J. Energy Res.*, 2007, **31**(13), 1275–1291.
7. K. Jorgensen, Technologies for electric, hybrid and hydrogen vehicles: Electricity from renewable energy sources in transport, *Utilities Policy*, 2008, **16**(2), 72–79.

8. H. Robert, *et al.*, Microgrid: A Conceptual Solution, *Proceedings of the 35th Annual IEEE Power Electronics Specialists Conference*, 2004, **6**, 4285–4290.
9. A. Carlos and A. Hernandez, Fuel Consumption Minimization of a Microgrid, *IEEE Trans. Indus. Appl.*, 2005, **41**(3), 673–681.
10. Y. Takuma and T. Goda, Microgrid for Urban Energy, *Trans. Soc. Heat. Air-Con. Sanitary Eng. Jpn.*, 2005, **79**(7), 573–579 (in Japanese).
11. H. K. Geyer, Dynamic response of steam-reformed, methanol-fueled, polymer electrolyte fuel cell systems, *31st Intersoc. Energy Convers. Eng. Conf.*, 1996, **2**, 1101–1106.
12. M. Ali, Development of Highly Efficient and Clean Engine System using Natural-Gas and Hydrogen Mixture Fuel Obtained from Onboard Reforming, *NEDO report ID:03B71006c*, 2005 (in Japanese).
13. D. Yap, Effect of Hydrogen Addition on Natural Gas HCCI Combustion, *SAE Tech. Pap. Ser.*, SAE-2004-01-1972, 2004.
14. J. F. Larsen and J. S. Wallace, Comparison of Emissions and Efficiency of a Turbocharged Lean-Burn Natural Gas and Hythane Fuelled Engine, *ASME ICE*, 1995, **24**, 31–40.
15. M. R. Swain and M. J. Usuf, The Effects of Hydrogen Addition on Natural Gas Engine Operation, *SAE Tech. Pap. Ser.*, SAE-932775, 1993.
16. S. Obara and K. Kudo. Study on Improvement in Efficiency of Partial Load Driving of Installing Fuel Cell Network with Water Electrolysis Operation, *Trans. Jpn. Soc. Mech. Eng., Series B,* 2005, **71**(701), 237–244 (in Japanese).
17. S. Obara and K. Kudo, Study of Efficiency Improvements in a Fuel-Cell-Powered Vehicle Using Water Electrolysis by Recovering Regeneration Energy and Avoiding Partial Load Operation, *Trans. ASME, J. Fuel Cell Sci. Technol.*, 2005, **2**, 202–207.

CHAPTER 2

Operation Analysis of a Compound Energy System – Exhaust Heat Use Plan when Connecting Solar Modules to a Fuel Cell Network

SHIN'YA OBARA

2.1 Introduction

In order to utilize effectively the power and the thermal energy produced in nuclear power plants and thermal power plants, it is necessary to reduce the transportation loss of energy. In particular, large heat release follows the transmission of thermal energy. Therefore, utilization of the exhaust heat in a distant place is not economical. From the viewpoint of energy transport, the distribution of a small energy-generation system is economical. This is because loss of energy transport decreases by installing an energy-generation system near the demand side. Furthermore, it is necessary to consider the introduction of small fuel cells and renewable energy from the viewpoint of the environment. In this chapter, the fuel system (hydrogen piping network), the electrical power system (power line network), and the heat-power system (hot-water piping network) of the fuel cells installed in each building are connected, and the best method of simultaneously generating two or more electrical power and heat loads of the buildings is examined. When two or more fuel cells are connected in a network, then the method involving cooperation and control of the

RSC Energy Series No. 3
Compound Energy Systems: Optimal Operation Methods
By Shin'ya Obara and Arif Hepbasli

Published by the Royal Society of Chemistry, www.rsc.org

electrical and heat power outputs is labeled a fuel-cell-energy network (FEN).[1,2] In an FEN, two or more fuel cells operated with a partial load with low efficiency can be stopped compulsorily. If fewer fuel cells are made to respond to these loads, decline in efficiency by a partial load may decrease.

The existing commercial power system can be used for a power network of FEN. Moreover, the existing town gas network can be used for a hydrogen piping network. However, a hot-water piping network must be newly installed. Since hot-water piping involves piping to all buildings, heat release is a problem. However, if a large-scale FEN is built in a cold district with considerable heat demand, the heat-release loss in a hot-water piping network would be predicted to represent the overall efficiency. When the paths of the hot-water piping differ, the amounts of heat released from the network also differ. Therefore, using the fuel cell exhaust heat effectively requires a path plan to minimize the amount of heat loss from the hot-water piping network. In this section, a program simulating the TSP analysis method (traveling salesman problem), using the genetic algorithm (GA) for the path plan of the hot-water piping network, is developed and investigated.[3–5] Since charging according to the volume of carbon dioxide gas discharged is being considered, connecting renewable energy and unused energy equipment to the FEN, which comprises tens of buildings, is also being investigated. For optimization using GA, there are previous reports that have explored the piping path and the equipment layout.[6–8] However, to the best of our knowledge, there have been no studies concerning the analysis of an energy network using fuel cells with a solar module accompanied by output change. Therefore, when connecting a solar module accompanied by output change to FEN, the path plan of the hot-water piping network, optimized to ensure minimal heat release, is investigated.

2.2 The Fuel Cell Energy Network with Solar Modules

2.2.1 Urban Area Model

An example where the hot-water piping network is applied to two or more buildings in Sapporo in Japan using the FEN is shown in Figure 2.1. S_1 to S_{12} in Figure 2.1 indicate the buildings in the network, while $L_{S_{i-(i+1)}}$ is the length of the hot-water piping connecting them. The urban area considered in the model is assumed to be Sapporo city in Japan, wherein the FEN is introduced as shown in Figure 2.1. The buildings include two-person households (TH), 3- or 4-person households (FH), 2 households living together with five or more persons (DH), a small-scale office (SO) and an apartment house (AP). The electrical power and thermal energy-demand patterns for each building in winter (February), summer (August), and mid-term (May) are shown in Figure 2.2.[9] The power capacity of the fuel cell installed in each building is decided at the maximum value of the power load of each building shown in Figure 2.2.

The average temperatures in Sapporo for the sampling time on representative days in February, May, and August are shown in Figure 2.3.[10] There is no

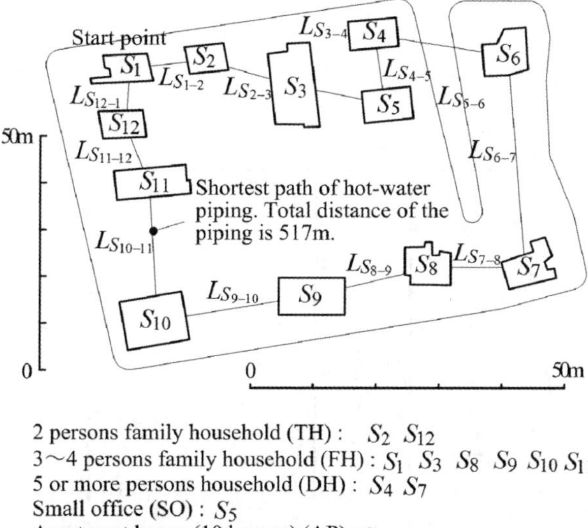

Figure 2.1 Fuel cell network model (FEN) installed in an urban area. The minimum length of the hot-water piping is 517 m.

cooling load during the summer in Sapporo. Electricity demand includes household appliances and electric lighting, and heat demand comes from heating, the hot-water supply, and a baths. The area of an average individual house (3- or 4-person household) in Sapporo is 140 m^2, with 2 storeys and the houses are made of wood.

2.2.2 Characteristic of the Solar Module

The power-generation area of a solar module is 24 m^2, and the maximum production of electricity is 3 kW. The experimental results when using this solar module in Sapporo City is shown in Figure 2.4.[11] The solar module is a single-crystal type and the installation angle is 60 degrees. The surface is fixed southward. Although it snows in winter (February), the modular surface is treated and snow can be further melted with an electric heater so that snow does not lie on the solar module. The production of electricity of the solar module influences the production of electricity of the fuel cell as described in Section 2.4. If the installation angle and capacity of the solar module are changed, the amount of exhaust heat of the fuel cell is affected. If the amount of exhaust heat of the fuel cell changes, the temperature of the hot water that recovers the exhaust heat also changes. After all, a change in the power output of a solar module influences the heat release of the hot-water piping.

In the analysis example of Section 2.4, the maximum electrical power output of the solar module installed in each building is rendered the same as the

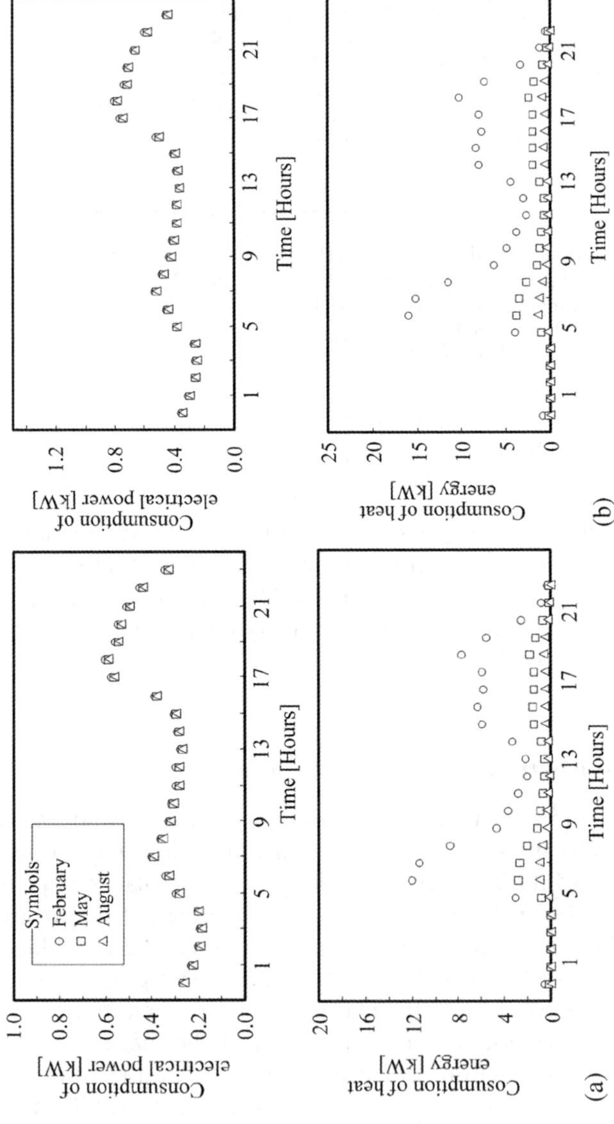

Figure 2.2 Energy-demand patterns; (a) TH: Family household (2 persons); (b) FH: Family household (3–4 persons); (c) DH: Family household (5 or more persons); (d) SO: Office; (e) AP: Apartment.

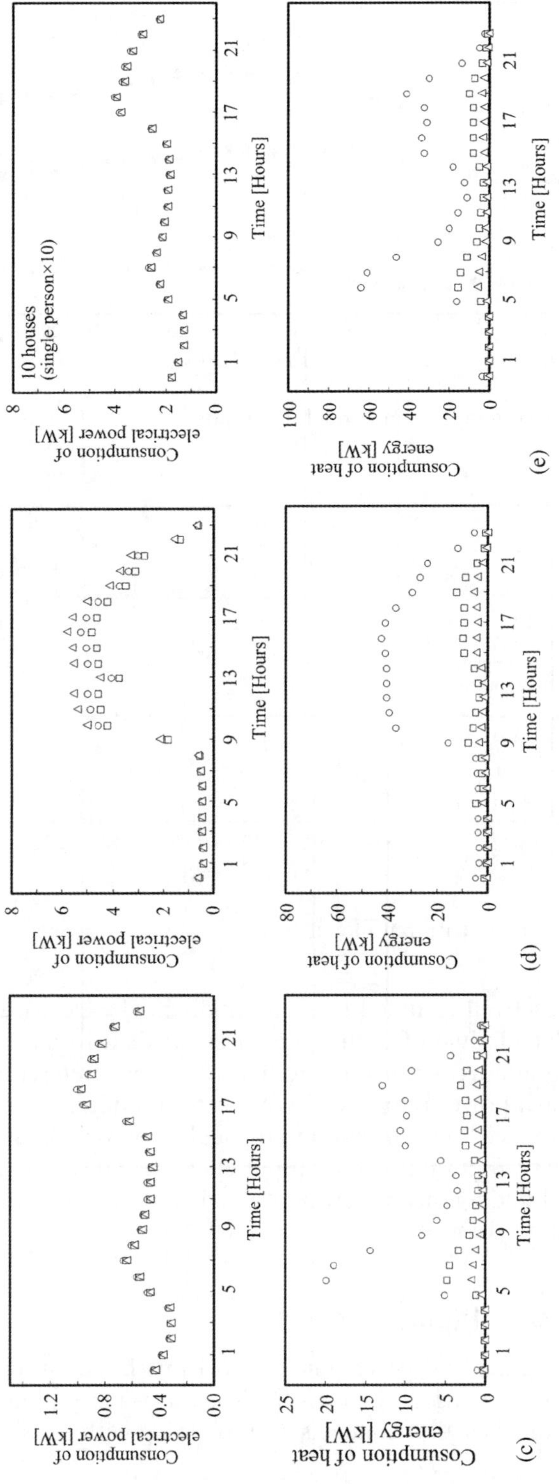

Figure 2.2 Continued.

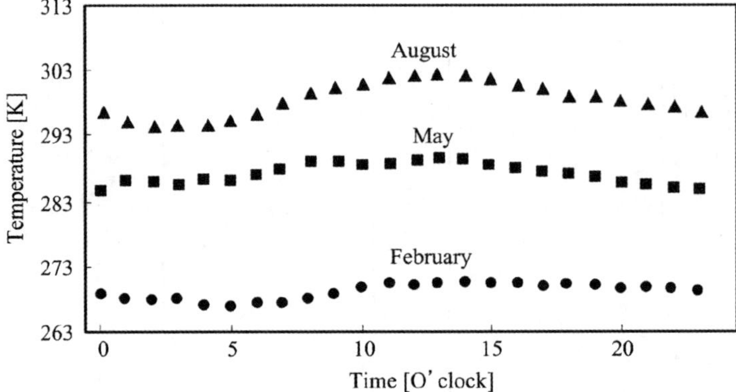

Figure 2.3 External temperature model in Sapporo.

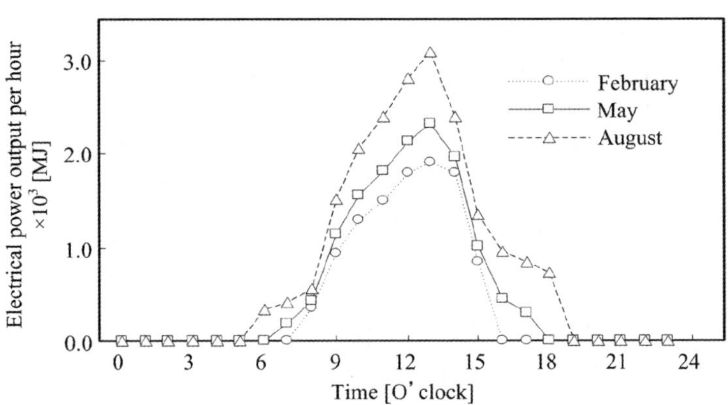

Figure 2.4 Solar module output model.

capacity of the fuel cell installed in each building. The electrical power generated by the solar module of each building is supplied to meet the electricity demand of the building, once this module has been preferentially installed. When the electrical power remains within a certain building, surplus electrical power can be supplied to each building through a power line network. In the analysis example of Section 2.4, the output characteristics of the solar module installed in each building mean they are treated relatively the same as shown in Figure 2.4.

2.2.3 Hot-Water Piping Network

Suppose reforming gas and power line networks within existing town gas and power line networks are used, respectively. The hot-water piping path shown in Figure 2.1 is of an order that connects each building with the shortest distance

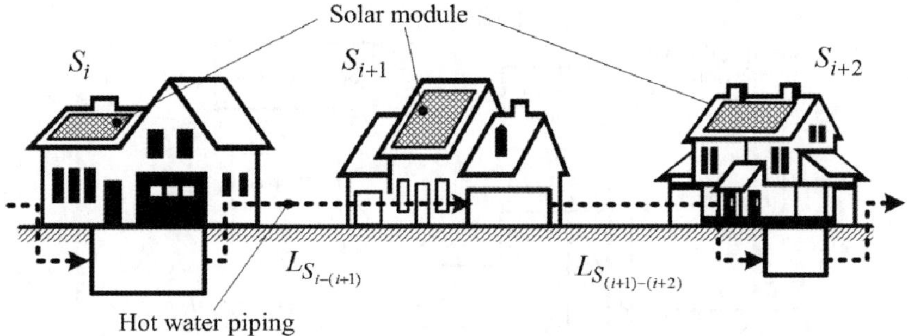

Figure 2.5 Hot-water piping network of FEN with solar modules.

(517 m). When the specification of the hot-water piping connecting each building are the same, assuming the heat release of the hot-water piping is dependent only on the piping length, the hot-water piping path intended to minimize heat release will always be the shortest path. The path of hot-water piping influences the heat release of the whole hot-water network. Details of the relationship between the piping path and heat release are presented in Section 2.3. Therefore, in order to use the exhaust heat or fuel cells effectively, a path plan of piping that minimizes the heat release in a hot-water piping network is required.

2.2.4 Facility Scheme

Figure 2.5 shows the model that connects hot-water piping to S_{i+2} from Building S_i. The heat output in the fuel cell installed in each building conveys the surplus and deficiencies through the hot-water piping, once the heat demand in the building has been satisfied. At first, after supplying the heat output from a fuel cell installed in a building to the heat demand of the building, excess heat (deficiency) is output (input) from the hot-water piping network. As shown in Figure 2.6(a), a fuel cell, solar module, inverter, and heat exchanger, able to output or input heat from or to the hot-water piping network, respectively, are installed in each building linked to the FEN. When the building in Figure 2.6(a) is set to S_i, it includes a hot-water input with a quantity of heat $H_{in,S_{i-1},t}$ at temperature $T_{in,S_i,t}$ from the hot-water piping network. Moreover, when the heat demand in S_i is set to $H_{S_i,need,t}$ and the exhaust heat output of a fuel cell is set to $H_{S_i,fc,t}$, the hot-water output from S_i is the quantity of heat $H_{out,S_i,t} = H_{in,S_{i-1},t} + (H_{S_i,fc,t} - H_{S_i,need,t})$ at temperature T_{out,S_i},t. The electrical power generated by a fuel cell is $E_{S_i,fc,t}$, that generated by the solar module is E_{S_i},sol,t, and that after changing into the regulation frequency of exchange with an inverter is $E_{S_i,need,t}$. The electrical power generated by a solar module is preferentially supplied to meet the electricity demand in the building currently installed. When electricity production E_{S_i},sol,t of a solar

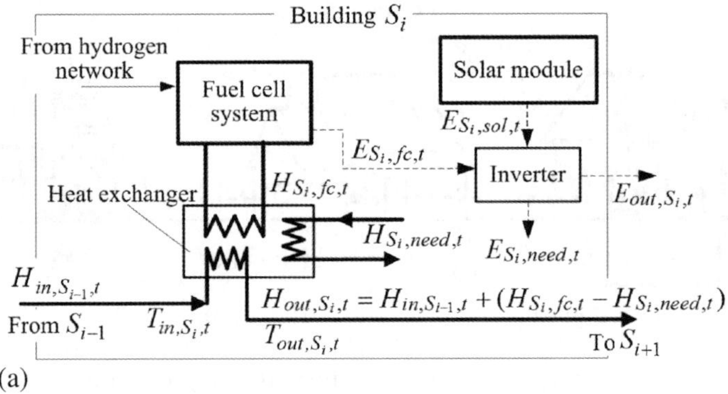

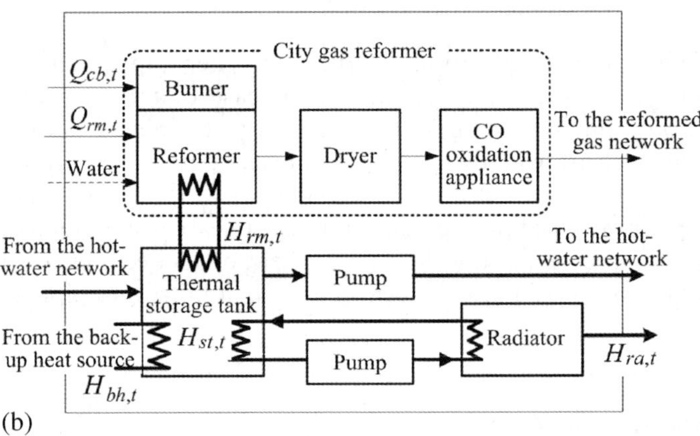

Figure 2.6 FEN system equipment; (a) Equipment installed in all buildings; (b) Common equipment of the FEN installed in the machinery room.

module is lacking and insufficient to meet the electricity demand $E_{S_i,need,t}$, this shortfall is compensated via the operation of a fuel cell ($E_{S_i,fc,t}$). Although the electrical power, excluding the electricity demand $E_{S_i,need,t}$ from E_{S_i},sol,t, can be supplied to any building in the FEN through the power line network, this is not taken into consideration in the case analysis described later.

Figure 2.6(b) presents a machinery room scheme showing the installation of common FEN facilities. The machinery room is assumed to be installed in a building linked to the FEN; the starting point of the hot-water piping network. A town-gas reformer, a heat-storage tank, a radiator, and backup heat source equipment are also installed there. Town gas, with a quantity of flow $Q_{rm,t}$, and water are supplied to a reformer, and reformed gas with a high hydrogen concentration is produced from the town gas through a steam reforming reaction using a catalyst. Steam reforming is an endoergic reaction and the heat source of this reaction is a supply of town gas of quantity of flow $Q_{cb,t}$ to a burner. The composition of hydrogen and steam in the reformed gas varies considerably. Carbon monoxide at a concentration of several per cent is typically present in the reformed gas when it comes out of the dryer. Therefore, CO oxidization is carried out using oxidation equipment so that the concentration of the carbon monoxide gas in reformed gas can be restricted to 10 ppm or less.

2.3 The Path Plan of a Hot-Water Piping Network

2.3.1 Heat-Transport Model of a Hot-Water Piping Network

Figure 2.7 shows the model about (a) the path of the FEN hot-water piping, (b) fuel cell capacities in each building, (c) electricity demand and the electrical power output of the fuel cells and solar modules, (d) change in the hot-water temperature, and (e) the heat-release volume in piping per unit length. In the model of Figure 2.7, a machinery room is installed in Building S_1, and the hot water for heat supply will flow in order of S_1 to S_7 as shown in Figure 2.7(a). As shown in Figure 2.7(b), the capacity of each fuel cell determines the scope for the maximum electrical power load in each building. The electricity generation from the fuel cell installed in a building is a value that affects the amounts of power output in a solar module relative to the amount of electricity demanded, as shown in Figure 2.7(c). Therefore, the amount of exhaust heat from the fuel cell installed in each building is influenced by the amount of electrical power output in a solar module. The fuel cell is installed in all the buildings through the FEN, and the outlet hot-water temperature of each building is decided; keeping in mind the balance relationship between the quantity of heat of the hot water input into each building, the exhaust heat volume of the fuel cell installed in the building, and the heat amount demand in the same. The hot-water quantity of the heat inputted into each building differs in terms of the path plan of hot-water piping. The outlet hot-water temperature of a building fluctuates on each building, as shown in Figure 2.7(d). Consequently, the heat-release volume per unit length of piping is subject to change as shown in Figure 2.7(e). Therefore, the amounts of heat released through the piping differ according to the building traversed by the hot-water piping itself. The path plan of the hot-water piping influences the system-wide energy efficiency.

Generally, the path of hot-water piping is decided by the shortest path length. Considering the characteristics of hot-water temperature, we should design the hot-water piping path of the distributed energy system other than by minimizing the radiating surface. This is because heat release differs in the profile of hot-water temperature, as Figures 2.7(d) and (e) show. The path of hot-water piping influences the heat release of the hot-water piping network.

2.3.2 Heat-Transfer Model of Hot-Water Piping

Figure 2.8 shows the model for heat input and output from the hot-water piping that connects Buildings S_i and S_{i+1}, and the heat-release volume in the piping is set to $H_{r,S_{i-(i+1)},t}$ in sampling time t. The hot-water temperature is $T_{in,S_i,t}$ and the quantity of heat $H_{in,S_i,t}$ is input into building S_i through the

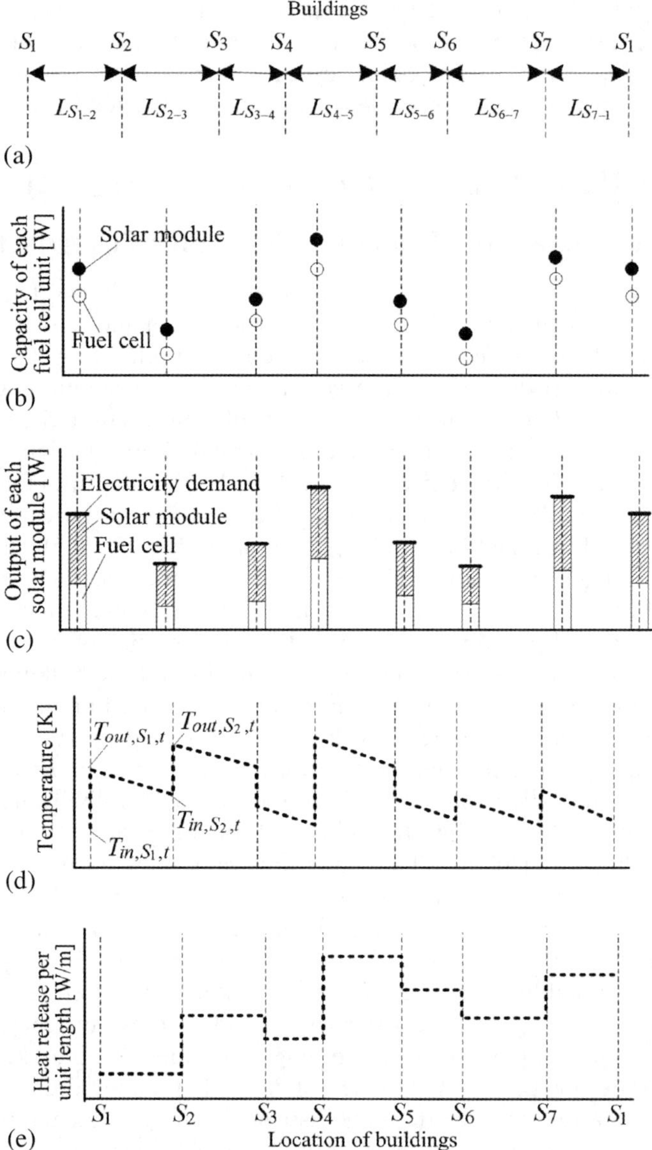

Figure 2.7 The allocation relationship between distributed fuel cells and heat release; (a) Route of hot-water supply; (b) Capacities of each fuel cell system and solar module; (c) Electricity demand and electrical power output; (d) Hot-water temperature; (e) Heat release of each hot-water piping.

network. When carrying out the power generation operation of the fuel cell installed in S_i so that the electricity demand $E_{S_i,need,t}$ may be satisfied, the exhaust heat output of the fuel cell is $H_{S_i,fc,t}$. Moreover, $H_{S_i,need,t}$ is the heat demand in S_i. The quantity of hot-water heat H_{out,S_i} output from S_i is

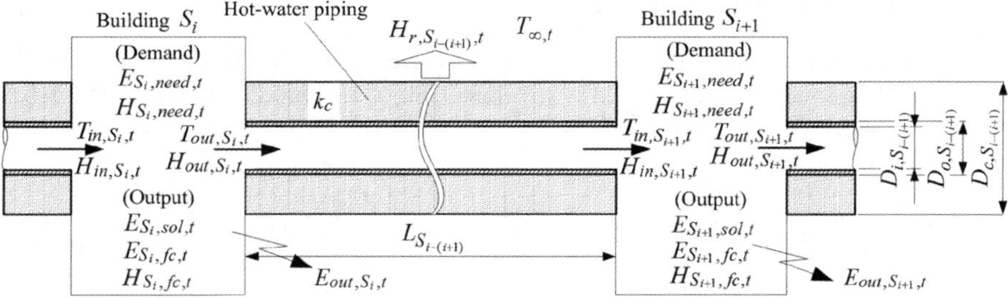

Figure 2.8 Hot-water piping model.

$H_{in,S_i,t} + H_{S_i,fc,t} - H_{S_i,need,t}$. Here, since the generation of electricity of a fuel cell $E_{S_i,fc,t}$ differs from the value of $E_{S_i,need,t}$, the values of $H_{S_i,fc,t}$ also differ. Section 2.3.4 describes the relation between the output of the electrical power and the heat of a fuel cell to the electrical power load. Although the temperature of the hot-water output from S_i is $T_{\mathrm{out},S_{i,t}}$, by the time the hot-water reaches S_{i+1}, there will be heat release $H_{r,S_{i-(i+1)},t}$ from the piping. Therefore, for the hot-water input into S_{i+1}, the temperature falls to $T_{in,S_{i+1},t}$ and the quantity of heat is $H_{in,S_{i+1},t} = (H_{out,S_i,t} - H_{r,S_{i-(i+1)},t})$. Furthermore, in Building S_{i+1}, since the power generation operation of the fuel cell is performed so that the electricity demand $E_{S_{i+1},need,t}$ is satisfied, the exhaust heat $E_{S_{i+1},fc,t}$ is output. This calculation is repeated for the following condition for all buildings: For a heat demand of $E_{S_{i+1},need,t}$ in S_{i+1}, the quantity of heat S_{i+1} of the hot-water output from $H_{out,S_{i+1},t}$ is $H_{in,S_{i+1},t} + H_{S_{i+1},fc,t} - H_{S_{i+1},need,t}$. Figure 2.8 shows the model of the hot-water piping connecting S_i and S_{i+1}. The bore diameter of the hot-water piping is expressed as $D_{i,S_{i-(i+1)}}$ the outside diameter is expressed as $D_{o,S_{i-(i+1)}}$, and the outside diameter of the heat-insulating material, with which the piping is equipped, is set to $D_{c,S_{i-(i+1)}}$. The heat conductivity of the piping material and thermal insulation are set to k_p and k_c, respectively. The coefficient of overall heat transmission $K_{S_{i-(i+1)}}$ between the hot-water and the surface of the heat-insulating material is determined as follows.

$$K_{S_{i-(i+1)}} = 1 \Bigg/ \left\{ \begin{array}{l} \dfrac{1}{h_{w,S_{i-(i+1)}} \cdot D_{i,S_{i-(i+1)}}} + \dfrac{1}{2 \cdot k_p} \ln \dfrac{D_{o,S_{i-(i+1)}}}{D_{i,S_{i-(i+1)}}} \\ + \dfrac{1}{2 \cdot k_c} \ln \dfrac{D_{c,S_{i-(i+1)}}}{D_{o,S_{i-(i+1)}}} + \dfrac{1}{h_{\infty,S_{i-(i+1)}} \cdot D_{c,S_{i-(i+1)}}} \end{array} \right\} \quad (2.1)$$

In the analysis example in this chapter, although the hot-water piping is installed above ground. $K_{S_{i-(i+1)}}$ can also be obtained by similar calculations for piping buried underground. The outlet hot-water temperature for Building S_i is $T_{out,S_i,t}$, and the outdoor air temperature in sampling time t is $T_{\infty,t}$. The heat-release volume $H_{r,S_{i-(i+1)},t}$ in the piping of length $L_{S_{i-(i+1)}}$ which connects S_i and S_{i+1} for this time is calculated by the following equation:

$$H_{r,S_{i-(i+1)},t} = K_{S_{i-(i+1)}} \cdot D_{c,S_{i-(i+1)}} \cdot \pi \cdot L_{S_{i-(i+1)}} \cdot (T_{out,S_i,t} - T_{\infty,t}) \quad (2.2)$$

2.3.3 Heat Energy Balance

Assuming that the buildings with fuel cells M are connected to the FEN and the fuel cells M generate electricity at sampling time t, the heat balance of the FEN is then expressed using

$$\sum_{i=1}^{M} H_{S_i,fc,t} + H_{rm,t} + H_{bh,t} + H_{st,t} = \sum_{i=1}^{M} H_{S_i,need,t} + \sum_{i=1}^{M} H_{r,S_{i-(i+1)},t} + H_{ra,t} \quad (2.3)$$

The left side of eqn (2.3) shows the heat outputs, and the right-hand side shows the heat consumption. The 1st term on the left side is the exhaust heat output by the fuel cell M, the 2nd term expresses the exhaust heat of a reformer, the 3rd term expresses heat supply from an auxiliary heat source, and the 4th term expresses the heat output from the heat-storage tank. The 1st term on the right-hand side is the heat demand in each building, the 2nd term is the heat-release volume by the hot-water piping that connects S_i and S_{i+1}, and the 3rd term expresses the heat-release volume in the radiator installed in the machinery room.

2.3.4 Analysis Method

(1) The output model of a fuel cell.

The relationship between the electrical power load ratio to the power-generation capacity of a fuel cell, defined as the load factor, and the ratio of heat output to the electrical power output, calculated based on the power-generation results of a proton-exchange membrane fuel cell (PEFC) stack, is shown in Figure 2.9(a).[1,2] This essentially represents the performance of the fuel cell stack. The characteristics of Figure 2.9(a) are the measurement results of the electrical power output at an AC–DC converter exit, and the heat output in a fuel cell stack exit. The performance was measured by supplying reformed gas and atmospheric air to the anode and cathode of a PEFC with a maximum output 10 W. An outline of the experimental device is shown in Figure 2.9(b). Output performance was measured by controlling the ambient temperature of the fuel cell. Exhaust heat was calculated from the mean specific heat and the rising rate of the representative temperature of the fuel cell. The average operating temperature of the PEFC model assumed in analysis is 353 K, and hydrogen fuel is obtained by the steam reforming of town gas. The load factor for the electrical power load $E_{S_i,need,t}$ of the fuel cell in Building S_i is calculated, and the heat output $H_{S_i,fc,t}$ of the fuel cell can be calculated using the plot in Figure 2.9(a).

(2) Capacity of a fuel cell.

The power-generation capacity of the fuel cell installed in each building is applied for 20% of margin, and the maximum power consumption in a whole year is roughly doubled by around 1.2, and then decided. The power-generation capacity of the fuel cell linked to FEN boosts the maximum annual power

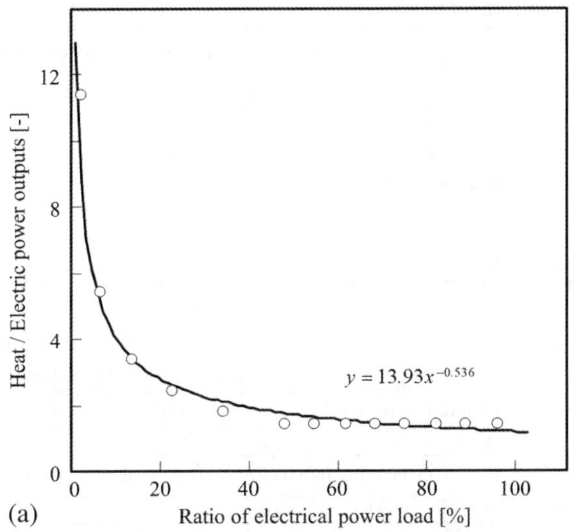

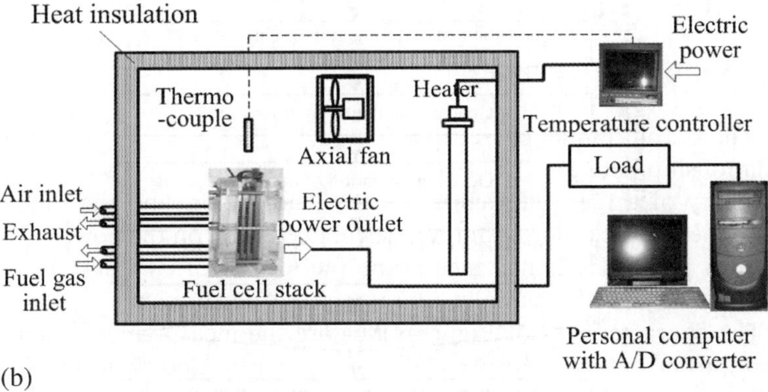

Figure 2.9 Fuel cell output model; (a) Fuel cell stack performance; (b) Fuel cell stack test system.

consumption value in each building by a factor of almost 1.2. The power-generation capacity of the fuel cell used in the analysis in the following section is shown in Table 2.1.

2.3.5 Analysis Flow

(1) The analysis program using the view of TSP by GA.

In this chapter, a program for the hot-water piping path using the traveling salesman problem (TSP) was developed.[5] The path of the hot-water piping is described by a chromosome model used by the genetic algorithm (Figure 2.10). Crossovers and mutations are added to many chromosome models, and an

Table 2.1 Fuel cell capacity of buildings.

Buildings	*Maximum electricity needs (kW)*	*Fuel cell capacity (kW)*
(a) TH: Family household (2 persons)	0.8	1.0
(b) FH: Family household (3–4 persons)	0.9	1.1
(c) DH: Family household (5 or more persons)	1.0	1.2
(d) SO: Office	6.0	7.5
(e) AP: Apartment (10 houses)	4.0	5.0

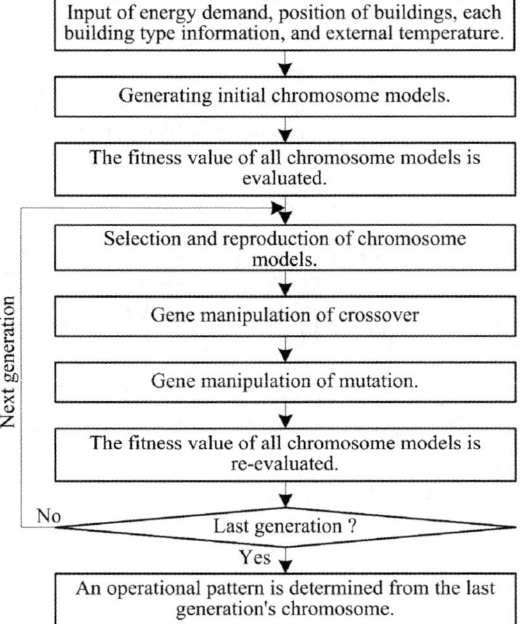

Figure 2.10 Analysis flow of the genetic algorithm.

analysis program evaluates the value of the objective function of each chromosome model. When the value of the objective function of a certain chromosome model fills the objective in a better fashion, this indicates that the "adaptive value" is high. In an analysis program, a chromosome model with a high adaptive value is made with a high probability of survival, and other chromosome models become extinct. A model with the highest adaptive value is the optimal solution for the chromosome models which are repeated and calculated for gene manipulation by crossovers and mutations of the chromosome models, and the evaluation and selection of adaptive value, and survive with the last generation number predecided. However, if the problem in this chapter is analyzed using a general GA, many chromosome models showing a path traversing the same building twice or more will be generated. For this reason, it is necessary to extinguish many chromosome models, and the calculation efficiency will thus significantly fall. Therefore, in this chapter, the view of order expression of the path by Dewdney is introduced. Since the chromosome model, which must be canceled by devising the order expression of a path, does not appear, the efficiency of the path planning analysis program improves.

(2) Objective function and adaptive value

A chromosome model with less heat release from the hot-water piping on a representative day is evaluated as an individual model with a high adaptive value. The objective function F is shown in the following equation:

$$F = \text{minimize}\left(\sum_{t=1}^{Day}\sum_{i=1}^{M} H_{r,S_{i-(i+1)},t}\right) \tag{2.4}$$

However, Period is the system operating period, M is the number of buildings connected to the FEN, and $H_{r,S_{i-(i+1)},t}$ is the heat-release volume in the hot-water piping at sampling time t and that connecting Buildings i and $i+1$ is expressed.

2.4 Case Study

2.4.1 Specifications of Hot-Water Piping

In the analysis example in this chapter, the hot-water piping used is assumed to be a hard vinyl pipe with an outside diameter $D_{o,S_{i-(i+1)}}$ of 50 mm, and a bore diameter $D_{i,S_{i-(i+1)}}$ of 40 mm, and the piping is assumed to be covered with a 30-mm thick polystyrene-foam system heat-insulating material ($D_{c,S_{i-(i+1)}}$ of 110 mm). Moreover, the heat-transfer coefficient h_w of the hot water in the piping is set to $3.5\,\mathrm{k\,W/m^2\,K}$, and the heat-transfer coefficient h_∞ between the heat-insulating-material surface and atmospheric air is $11.6\,\mathrm{W/m^2\,K}$. These values are assigned to eqn (2.2) and consequently, $K_{S,t}$ is calculated to be $0.05\,\mathrm{W/m^2\,K}$.

2.4.2 Analysis Procedure

The analysis procedure involved in planning the path of a hot-water piping network is described below. First, the coordinates of each ridge of the urban area model shown in Figure 2.1, the energy-demand pattern of the building shown in Figure 2.2, and the outside temperature data of the representative day for each month in Sapporo shown in Figure 2.3, are input into the program developed in this chapter. Next, N_{cr} chromosome models showing the path order of the hot-water piping are prepared at random. For each of these chromosome models, the objective function shown by eqn (2.4) is evaluated. The top R_{cr}% of the high chromosome model of the adaptive value is made to survive, and the remaining chromosome models are excluded. Crossover and mutations are added to the surviving chromosome models using the probabilities r_{cs} and r_{mu}.

Furthermore, predecided N_{ge} generation numbers repeat the calculation to evaluate the adaptive value of the chromosome models and selecting individuals with a low adaptive value. The hot-water piping path, obtained for the highest adaptive0value individual, is made into an optimal path in the previous generation's chromosome models. In the calculation of eqn (2.4), the heat $H_{r,S_{i-(i+1)},t}$ released in the piping, which connects each building in the equation, is calculated using the following procedures. From the electricity demand pattern in Building i, the electricity demand $E_{S_i,need,t}$ in the sampling time t is found. The power-generation capacities of the fuel cells installed in each building are shown in Table 2.1, and the characteristics of electrical power and heat output are obtained from Figure 2.9. The load factor of the electrical power can then be calculated by the power-generation capacity of a fuel cell, and $H_{S_i,fc,t}$ can be obtained by applying the load factor and $E_{S_i,need,t}$ to the curve in Figure 2.9. The heat storage capacity is predecided, and when the heat input exceeds this capacity, $H_{ra,t}$ of the quantity concerned is assumed to be emitted from a radiator. The heat-release volume from the hot-water piping ($H_{r,S_{i-(i+1)}}t$) within the sampling time t can be calculated based on the ability to give $H_{S_i,fc,t}$, $H_{rm,t}$, $H_{bh,t}$, $H_{St,t}$, $H_{S_i,need,t}$, and $H_{ra,t}$ to eqn (2.3).

2.4.3 Analysis Conditions and Parameters

The number N_{cr} of the chromosome models is 3000, and R_{cr}=40% are made to survive in the high order of adaptive value during selection. However, a randomly chosen chromosome model is generated and added instead of the 60%, which became extinct. Based on the results of maintaining chromosome model diversity, and trial and error, the crossover probability r_{cs} is assigned to be 70%, and the mutation probability r_{mu} assigned to be 1%. The generation number N_{ge} is 100. Using these parameters, two or more analyses are conducted, the highest chromosome model of adaptive value is discovered, and the path of the hot-water piping, the building used as a starting point, the flow direction of the hot water, and the heat-release volume on the representative day are, respectively, determined. However, the exhaust heat temperature of a fuel cell is set to 353 K, and the temperature of the hot-water that returns from the hot-water piping network to the machinery room is assumed to be nearly 333 K.

2.5 Analysis Results

2.5.1 Results of the Hot-Water Piping Path in FEN that Does not Connect Solar Modules

Figure 2.11 shows the results of searching for the FEN hot-water piping path in the case where a solar module is not connected, using the city area model

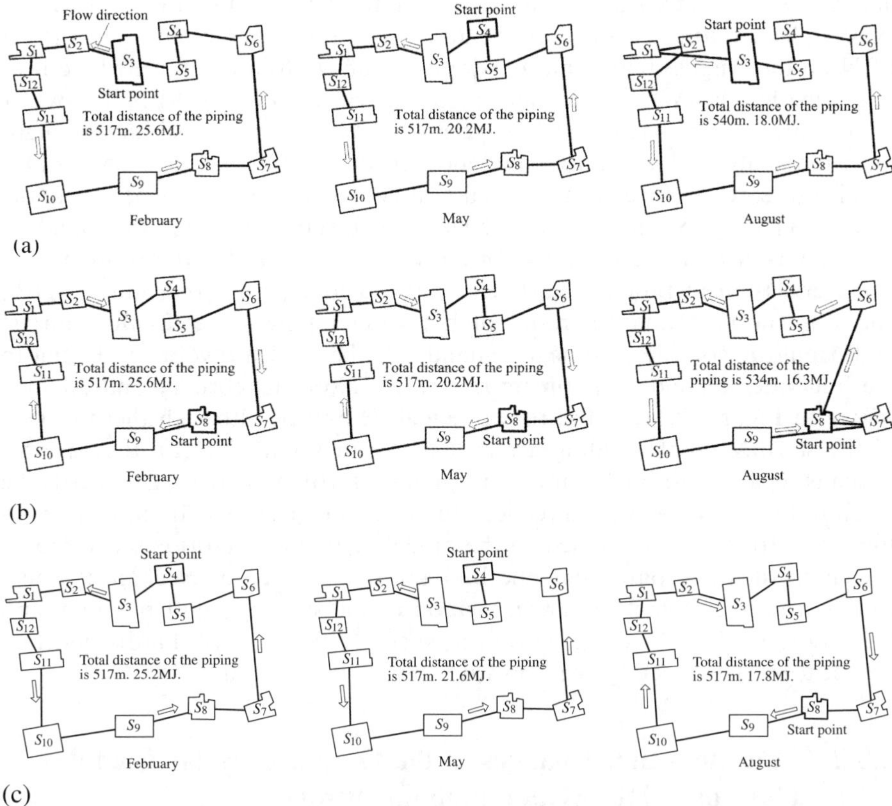

Figure 2.11 Results of the optimization analysis of a hot-water piping path without a solar module. (a) The energy-demand pattern of all buildings is based on a 2-person family household (TH). (b) The energy-demand pattern of all buildings excluding S_5 is based on a 2-person family household (TH). The energy-demand pattern of S_5 is the office (SO). (c) The building arrangement is the pattern shown in Figure 2.1.

shown in Figure 2.1. However, in order to evaluate search results, let the energy-demand pattern of all the buildings be a two-person household (TH) in Figure 2.11(a), the building S_5 of the model of Figure 2.11(a) is set as a small-scale office (SO) in Figure 2.11(b), and each energy-demand pattern shown in Figure 2.1 is used in Figure 2.11(c). The path of the hot-water piping, its length, the starting point building, the flow direction of the hot water, and the heat-release volume on the representative day are shown in each part of Figure 2.11. When the energy-demand pattern of all buildings shown in Figure 2.11(a) is TH, the hot-water piping path on representative days in February and May is similar to the shortest path or that shown in Figure 2.1. However, the results on the August representative day showed the path of Buildings S_1 and S_2 was not the shortest distance. This is because the heat amount demanded for the building connected to FEN on a representative day in August was extremely small compared to the quantity of the fuel cell exhaust heat outputted on each building. The hot-water piping network of FEN on the August representative day was subject to a situation where heat was superfluous. When the path planning calculation of FEN hot-water piping takes place when heat is superfluous, there is no possibility of high-precision analysis. The reduced analytical accuracy due to surplus heat is also visible on representative days in August, as shown in Figure 2.11(b), where the path of Buildings S_7 and S_8 is not the shortest distance. On the other hand, in the analysis of Figure 2.11(c), two or more energy-demand patterns are mixed, and there is considerable network heat demand compared with Figures 2.11(a) and (b). The path analysis result of hot-water piping serves as the shortest path from the considerable heat demand. The heat-release volume relating to the hot-water piping paths on representative days in February and August, shown in Figures 2.11(a) and (b) are equal. However, although the positions of the starting point building are S_3 and S_4; in Figure 2.11(a), they are S_8, which is separated from S_3 and S_4 in Figure 2.11(b). Moreover, as regards the result of the hot-water flow direction, Figures 2.11(a) and (b) are in reverse. In Figure 2.11(b), the building S_5 of the model of Figure 2.11(a) is set to SO. If the energy-demand pattern of the building is linked to FEN changes, there will be a visible difference in the position of the starting point building, and the hot-water flow direction. However, such a difference will hardly generate any heat release affecting the hot-water piping or its path.

2.5.2 Influences that Changes in the Output of Solar Modules Have on a Hot-Water Piping Network

(1) When all energy-demand patterns are the same.

Figure 2.12 shows the results when analyzing the energy-demand pattern of all buildings as two-person households (TH). Figure 2.12(a) shows the calculation results using the model in the solar module installed in Sapporo; as shown in Figure 2.4. The outputs of a solar module are based on the results of calculating Figure 2.12(b) at 80%, and Figure 2.12(c) at 60%, respectively. Moreover, Figure 2.13 shows the calculation result of the heat amount demanded on the representative day each month, with the heat-release volume for the whole hot-water piping network shown in Figure 2.12. If the solar modules installed in each building generate considerable quantities, that generated by each fuel cell will decrease relatively. Consequently, if the quantity generated by a solar module increases, the exhaust heat volume of a fuel cell will also decrease. However, when the electrical power load ratio of a fuel cell is less than 10%, based on the properties shown in Figure 2.4, the exhaust heat output to an electrical power output is large. The quantity generated by the fuel cell of each building in Figure 2.12 is the value where the quantity generated by the solar module based on the analysis result of Figure 2.11(a) is deducted. Therefore, so much heat in the shape of hot water flows through a hot-water piping network that generation by the solar module is scarce. Although the hot-

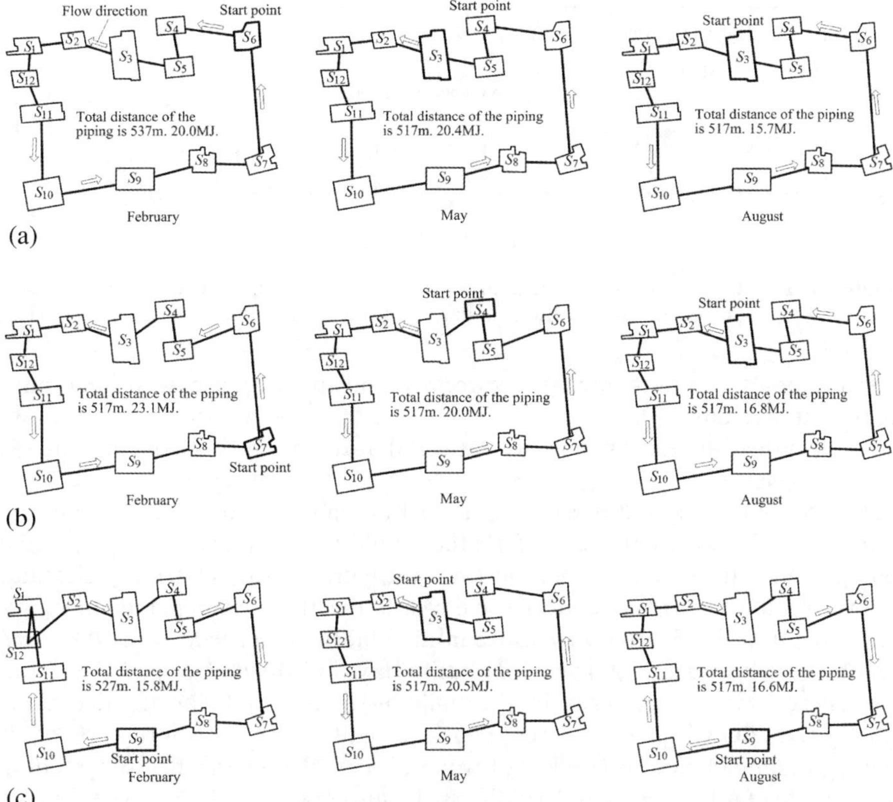

Figure 2.12 Results of the optimization analysis of a hot-water piping path with a solar module. The energy-demand pattern of all buildings is based on a 2-person family household (TH). (a) The power generation of the solar module uses 100% of a standard electric power output pattern; (b) The power generation of the solar module uses 80% of a standard electric power output pattern; (c) The power generation of the solar module uses 60% of a standard electric power output pattern.

water quantity of heat flowing through the hot-water piping network is small, of the order of Figure 2.11(a) and Figure 2.12, the heat-release volume of a hot-water piping network does not serve as this order; as shown in Figure 2.13. The reason is (a) as eqn (2.2) showed, the heat-release volume of a hot-water piping network is related to the outside air temperature, and (b) the period generated by a solar module is restricted to the time from early morning to evening. In particular, the heat amount demanded in TH includes more time zones previous and after that on representative days in February. Therefore, the heat-release volume in a hot-water piping network is decided not just based on the quantity generated by a solar module. Consequently, Figures 2.12(a)–(c) show the analysis results of the path of hot-water piping, the building used as a

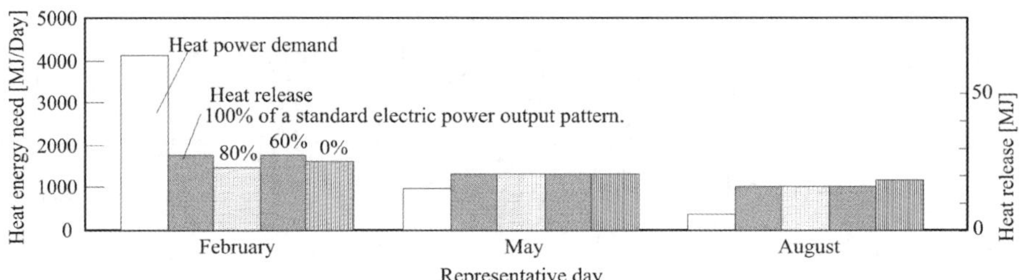

Figure 2.13 Heat energy demand and heat-release volume. The energy-demand pattern of all buildings is based on a 2-person family household (TH).

starting point, the hot-water flow direction and the heat-release volume on a representative day.

(2) When the energy-demand pattern that differs for only one building is given.

Figure 2.14 shows the results of setting Building S_5 into SO, and analyzing the energy-demand pattern of all other buildings included as TH. As the energy-demand pattern of SO shows in Figure 2.2(d), electricity demand between 9:00 to 18:00 is considerable. Although the heat demand on representative days in February is also considerable between 9:00 to 21:00, there will be little heat demand on representative days in May and August. When all the energy-demand patterns in the building linked to FEN are the same, analysis results (Figure 2.12) represent the path where the shortest or other corresponding choice applies to hot-water piping. However, when the building whose energy-demand pattern differs from other buildings extremely is installed into Building S_5, as shown in Figure 2.14, the path analysis results of hot-water piping differ in Figure 2.12. The optimal path of the hot-water piping network in FEN is considered to be influenced by the generation capacity of the fuel cells of each building, the quantity generated by solar modules, and energy-demand patterns. If a large scale fuel cell is installed in FEN, a different solution from the results of the analysis shown in Figure 2.14 should be obtained. Therefore, it is thought that the heat-release volume of the hot-water piping previous and subsequent to such a building with a large fuel cell capacity will exceed that of the hot-water piping connecting other buildings. The path planning calculation of hot-water piping of parts with relatively low heat-release volumes requires higher analytical accuracy compared with hot-water piping parts with considerable heat release. The complex path results for hot-water piping over every month of Figure 2.14(a), February and August of Figure 2.14(b) and February and August of Figure 2.14(c) indicates analytical accuracy as the cause. Figure 2.15 shows the calculation result of the heat amount demanded on representative days, and the heat-release volume in the whole hot-water piping network shown in Figure 2.14 every month. By having installed SO with much energy demand in FEN compared with TH, there are large amounts of heat release in hot-water piping in each month representative day.

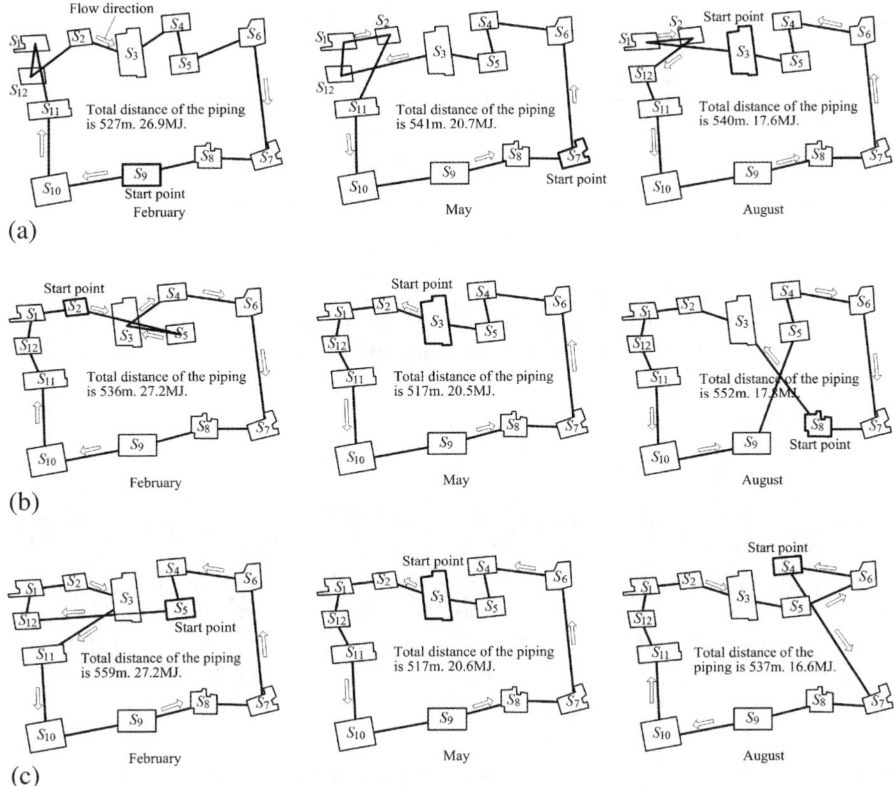

Figure 2.14 Results of the optimization analysis of a hot-water piping path. The energy-demand pattern of all buildings excluding S_5 is based on a 2-person family household (TH). The energy-demand pattern of S_5 is the office (SO). (a) The power generation of the solar module uses 100% of a standard electric power output pattern; (b) The power generation of the solar module uses 80% of a standard electric power output pattern; (c) The power generation of the solar module uses 60% of a standard electric power output pattern.

(3) In the case of the building model of Sapporo city.

Figure 2.16 shows the results of analyzing hot-water piping paths using each energy-demand pattern shown in Figure 2.1. As for SO of Building S_5, and AP of Building S_6, there is much energy demanded in the analysis of Figure 2.16. If there is a building that has a far greater fuel cell capacity than the FEN neighborhood, as described in the preceding paragraph, the analytical accuracy in the building neighborhood with low energy consumed will be poor, and, unlike the shortest path, often becomes complex. As the results of the hot-water path in May of Figure 2.16(a), February and August of Figure 2.16(b) and May and August of Figure 2.16(c) shows under the influence of Building S_5 and S_6, the piping path may not be the shortest. Figure 2.17 shows the calculation results of the heat amount demanded on representative days, and the

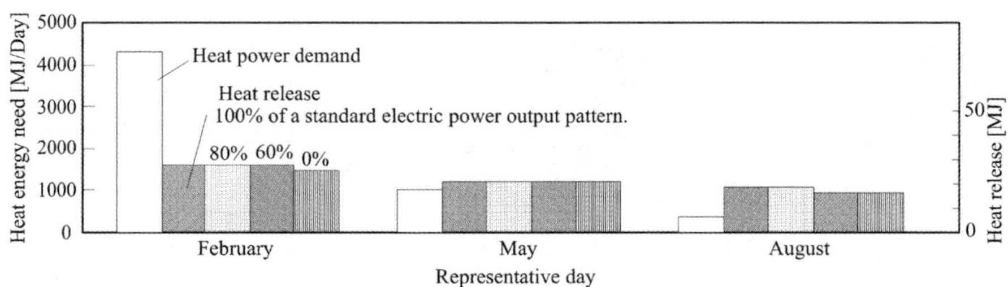

Figure 2.15 Heat energy demand and heat-release volume. The energy-demand pattern of all buildings excluding S_5 are those with 2-person family households (TH). The energy-demand pattern of S_5 is the office (SO).

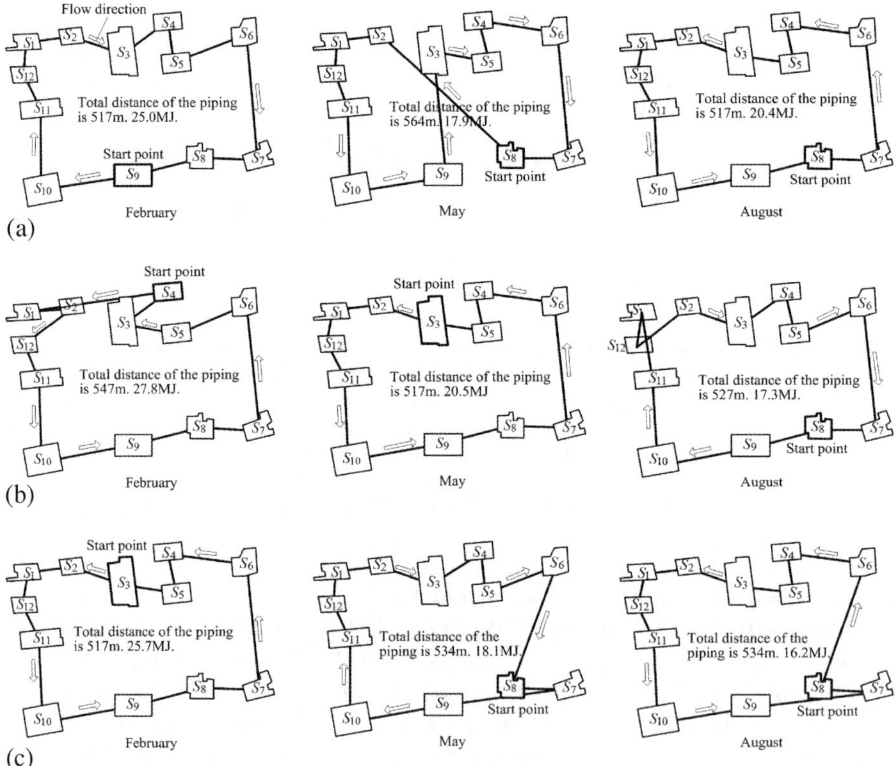

Figure 2.16 Results of the optimization analysis of a hot-water piping path. The building arrangement is the pattern shown in Figure 2.1. (a) The power generation of the solar module uses 100% of a standard electric power output pattern; (b) The power generation of the solar module uses 80% of a standard electric power output pattern; (c) The power generation of the solar module uses 60% of a standard electric power output pattern.

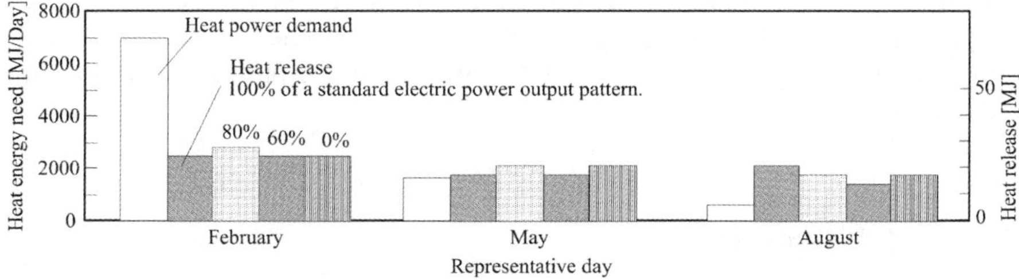

Figure 2.17 Results of the optimization analysis of a hot-water piping path with a solar module. The building arrangement is the pattern shown in Figure 2.1.

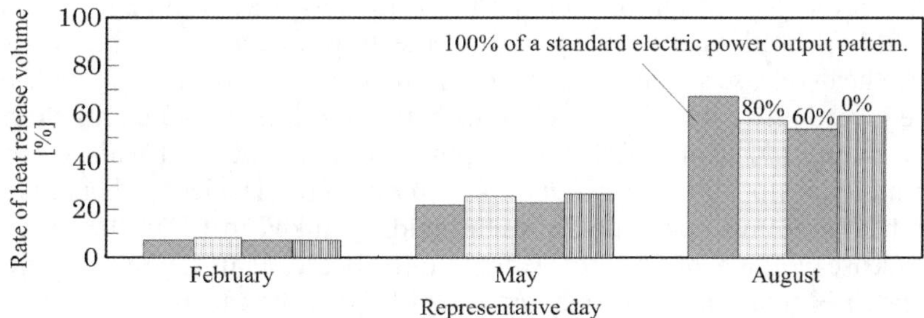

Figure 2.18 Rate of heat-release volume. The building arrangement is the pattern shown in Figure 2.1.

heat-release volume in the whole hot-water piping network shown in Figures 2.16 and 2.18 shows the heat-release volume rate of a hot-water piping network based against the heat amount demanded on representative days every month. The outside air temperature on representative days in February is low compared with that in August. As for the heat-release volume when floating the same quantity of heat for hot-water piping, February is larger. However, based on the analytical results of Figure 2.18, concerning the heat release rate of the hot-water piping network occupied with the heat demand on representative days each month, February is the smallest and August the largest. The difference in the outside air temperature of the area in which FEN is installed is not thought to influence the heat-release volume in hot-water piping strongly; based on the analytical results of Figure 2.18. Therefore, if considerable heat amounts are demanded by each building linked to FEN, the rate of the heat-release volume compared to the heat amount demanded is small, and the exhaust heat of fuel cells can be used effectively. Furthermore, the plan to change a hot-water piping path according to the season, and to minimize the heat-release volume of a hot-water network is meaningless in terms of the effect and expense required. However, changing the flow direction of the hot water has a small role to play in the heat-release volume in a hot-water network. In order to realize the effective use of fuel cell exhaust heat, fuel cell capacities and the capacity of solar modules are determined from the energy-demand pattern in each building linked to FEN, and after predicting the output range of the solar modules, it is further necessary to implement the path plan for the hot-water piping.

2.6 Conclusions

An effective method of using fuel cell exhaust heat was considered with reference to the fuel cell energy network (FEN) that operates fuel cells installed in two or more buildings synergistically. In this chapter, an analysis program to deal with the path-plan problem when supplying fuel cells exhaust heat to each building linked to FEN with hot-water to GA was developed and investigated. In the developed program, the path of the hot-water piping, its length, the building used as a starting points, the flow direction of the hot-water, and the heat-release volume of a representative day can be investigated. Based on the investigative results, the relations between the energy-demand pattern in each building linked to FEN, the output volume of the electrical power of a solar module, and the hot-water piping path were clarified. When there are many heat amounts demanded in each building linked to FEN, the rate of the heat-release volume of the hot-water piping based against the heat amount demanded is small, and the exhaust heat of fuel cells can be effectively used. Furthermore, considering the effective use of the fuel cell exhaust heat linked to FEN, it is effective in determining the fuel cell capacity and the capacity of solar modules based on the energy-demand pattern in each building, to predict the output range of a solar module, and to implement the path plan for the hot-water piping.

Acknowledgments

This work was partially supported by the Grant-in-Aid for Scientific Research(C) from the JSPS.KAKENHI(17510078).

Nomenclature

D_c: outside diameter of an insulating material m
D_i: bore diameter of the hot-water piping m
D_o: outside diameter of the hot-water piping m
E: electrical power W
E_{sys}: electrical power of an inverter outlet W
F: objective function J
H: heat power W
H_r: heat release from the hot-water piping W
h_w: heat-transfer coefficient between the hot water and the piping surface $W/m^2\,K$

h_∞: heat-transfer coefficient between the atmosphere and the heat-insulating-material surface W/m^2 K
K: coefficient of the overall heat transmission W/m^2 K
k_c: thermal conductivity of the heat-insulating-material W/m K
k_p: thermal conductivity of the piping W/m K
L: length of the hot-water piping m
M: fuel cell installation number
N_{cr}: number of individual chromosome models
N_{ge}: generation number
Q: quantity of flow m^3/s
R_{cr}: probability of survival %
r_{cs}: probability of mutation %
r_{mu}: probability of crossover %
S_i: the number of a building
T: temperature K
T_∞: external air temperature K
t: sampling time s

Subscripts

bh: back-up heat source
cb: burner
fc: fuel cell
in: input
need: energy need
out: output
ra: radiator
rm: reformer
sol: solar module
st: heat-storage tank

The names of buildings

AP: apartment house
TH: family household (2 persons)
DH: two-household house (5 or more persons)
FH: family household (3–4 persons)
SO: small office

References

1. S. Obara and K. Kudo, Study on Small-Scale Fuel Cell Cogeneration System with Methanol Steam Reforming Considering Partial Load and Load Fluctuation, *Trans. ASME, J. Energy Res. Technol.*, 2005, **127**, 265–271.

2. S. Obara, K. Kudo, Route Planning of Heat Supply Piping of Fuel Cell Energy Network. *Trans. Jpn. Soc. Mech. Eng.*, 2005, **71**(704)B, 1169–1176 (in Japanese).
3. D. E. Goldberg, Genetic Algorithms in Search, *Optim. Mach. Learning*, 1989, Addison Wesley.
4. J. E. Baker, Adaptive Selection Methods for Genetic Algorithms, *Proc. 1st Int. Joint Conf. on Genetic Algorithms*, 1985, ICGA85, 101–111.
5. A. K. Dewdney, Exploring the field of genetic algorithms in a primordial computer sea full of flobs, *Sci. Am.*, 1985, **85**(5), 16–21.
6. Y. Hongmei, F. Haipeng, Y. Pingjing and Y. Yi, A Combined Genetic Algorithm/Simulated Annealing Algorithm for Large Scale System Energy Integration, *Comput. Chem. Eng.*, 2000, **24**, 2023–2035.
7. Z. David, C. L. Jean, *Proceedings of the 1991 IEEE International Conference on Robotics and Automation*, 1991, 1940.
8. S. K. Sang, M. Sehyun and H. H. Soon, *J. Ship Prod.*, 1999, **15**(1), 1.
9. O. Nagase, T. Ikaga, T. Chikamoto, Study on the Effect of Energy Saving Methods and Global Warming Prevention in Tokyo, *19th Energy System-Economics-Environment Conference*, 2003, 461–466 (in Japanese).
10. National Astronomical Observatory, Rika Nenpyo, *Chronological Scientific Table CD-ROM*, 2003, Maruzen Co., Ltd.
11. K. Nagano, Feasibility study of the solar energy power-generation system in Sapporo, Tomita, K. (eds), *Proceedings of 33rd SHASE Hokkaido Affiliate Conference*, 1999, 5–8 (in Japanese).

CHAPTER 3

Operation of Compound Energy System – Fuel Cell Network System Considering Reduction in Fuel Cell Capacity

SHIN'YA OBARA

3.1 Introduction

In order for installation of the fuel cell system to houses or a small-scale and mid-scale building to spread, it is necessary to reduce equipment cost. Consequently, a fuel system network (hydrogen piping and oxygen piping) and an energy network (a power transmission line and hot-water piping) of distribution fuel cells are proposed.[1] In this system, common auxiliary machinery is installed in a machinery room. In this chapter, in order to reduce the capacity of the fuel cell connected to the network, a method of leveling the load is proposed. By this method, hydrogen and oxygen are generated by water electrolysis at times of low load with little power demand, and each gas is compressed and stored. On the other hand, the stored gas is supplied and generated to the fuel cell in a period of large power load. The experimental result shows that the power-generation characteristics improve greatly compared with air supply, when supplying oxygen to the fuel cell.[2] Therefore, if the oxygen generated when the load is small can be used for a high-load period, the installed capacity of the fuel cell can be reduced. Moreover, the heat-energy network is hot-water piping, and supplies heat to each building. Hot-water piping distributes heat via each building. When there is heat excess with some buildings, it can also recover

RSC Energy Series No. 3
Compound Energy Systems: Optimal Operation Methods
By Shin'ya Obara and Arif Hepbasli

Published by the Royal Society of Chemistry, www.rsc.org

this heat through the hot-water piping. In a heat-energy network, the hot-water temperature in a building outlet changes with the heat consumed by each building and the fuel cell exhaust heat of each building. Therefore, the heat release of the overall network differs according to the outside air temperature, piping distance, the starting point of the hot-water supply, and the flow direction of the hot water. Consequently, to counteract the piping heat-release loss of the heat-energy network, the minimum piping route is examined.

In the analysis case, the capacity-reduction effect of a fuel cell when installing load leveling using the water electrolyzer described above is investigated regarding a local energy network that includes houses, a hospital, a factory, an office, and a convenience store. Furthermore, the hot-water piping route and the fuel cell capacity placed on each building when optimizing the system with the object of minimizing the hot-water piping heat release are considered.

3.2 Load Leveling and Arrangement Plan of Fuel Cell

3.2.1 Fuel Cell Network System

The network model with the proton-exchange membrane fuel cell (PEFC) installed that is assumed in this chapter is shown in Figures 3.1 and 3.2. As shown in each figure, the fuel system (hydrogen piping and oxygen piping), the power system (power transmission line), and the heat-energy network (hot-water piping) between the fuel cells installed in each building are connected. A heat-transfer medium (hot water) is flowed in hot-water piping, the exhaust heat of a fuel cell is recovered, and heat is distributed to each building. The route of the hot-water piping can be set up arbitrarily, and the flow direction is one way, as the arrow in each figure shows. Figure 3.1 shows the system that supplies the power to Buildings A to G from one fuel cell stack installed in the machinery room, and is described as the R1 type below. A machinery room can be installed in an arbitrary building (Building A in Figure 3.1). As shown in Figure 3.1(c), the fuel cell (1: this number corresponds to that in Figure 3.1), water electrolyzer (6), city-gas reformer (7), hydrogen and oxygen compressor (9 and 11), cylinders (10 and 12), geothermal heat pump (13), heat-storage tank (14), *etc.* are installed in the machinery room. The heat output by fuel cell exhaust heat, the heat-storage tank, and the geothermal heat pump is distributed to each building through a heat-transfer medium. The piping route can be planned arbitrarily and it is in the order of Building ABCFEDGA in the example of Figure 3.1(a). As shown in Figure 3.1(b), headers (4 and 5) are set in each building at a hot water gate. The heat of the radiator (3) and a heat exchanger connected to the header is used for space heating and hot-water supply. Figure 3.2 shows the system that distributes a fuel cell in all the buildings, and this system is described as the R2 type below. Although the number of fuel cells increases and the equipment cost increases for the R2 type, heat-release loss with heat transport is small. The hot-water piping route of the

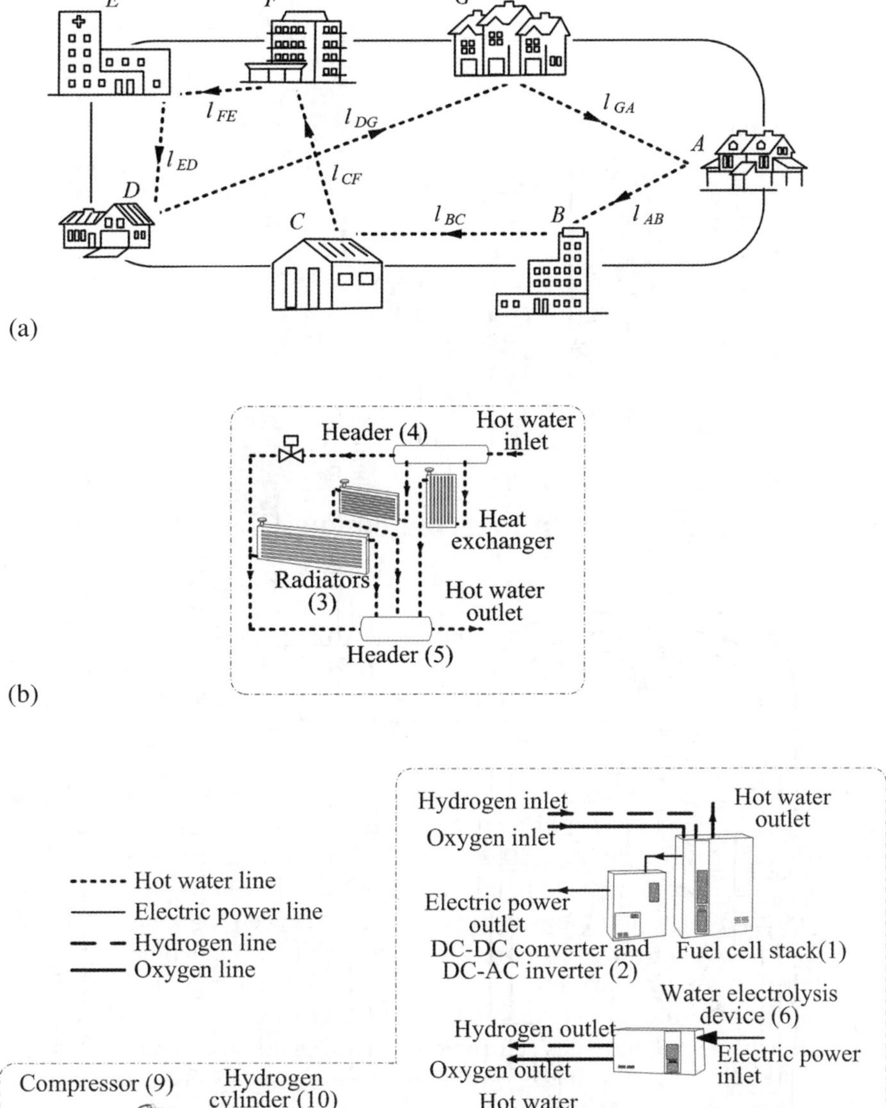

Figure 3.1 Fuel cell network system model (R1 type); (a) Fuel cell network model; (b) Radiator unit installed in building; (c) Auxiliary machine installed in utility room.

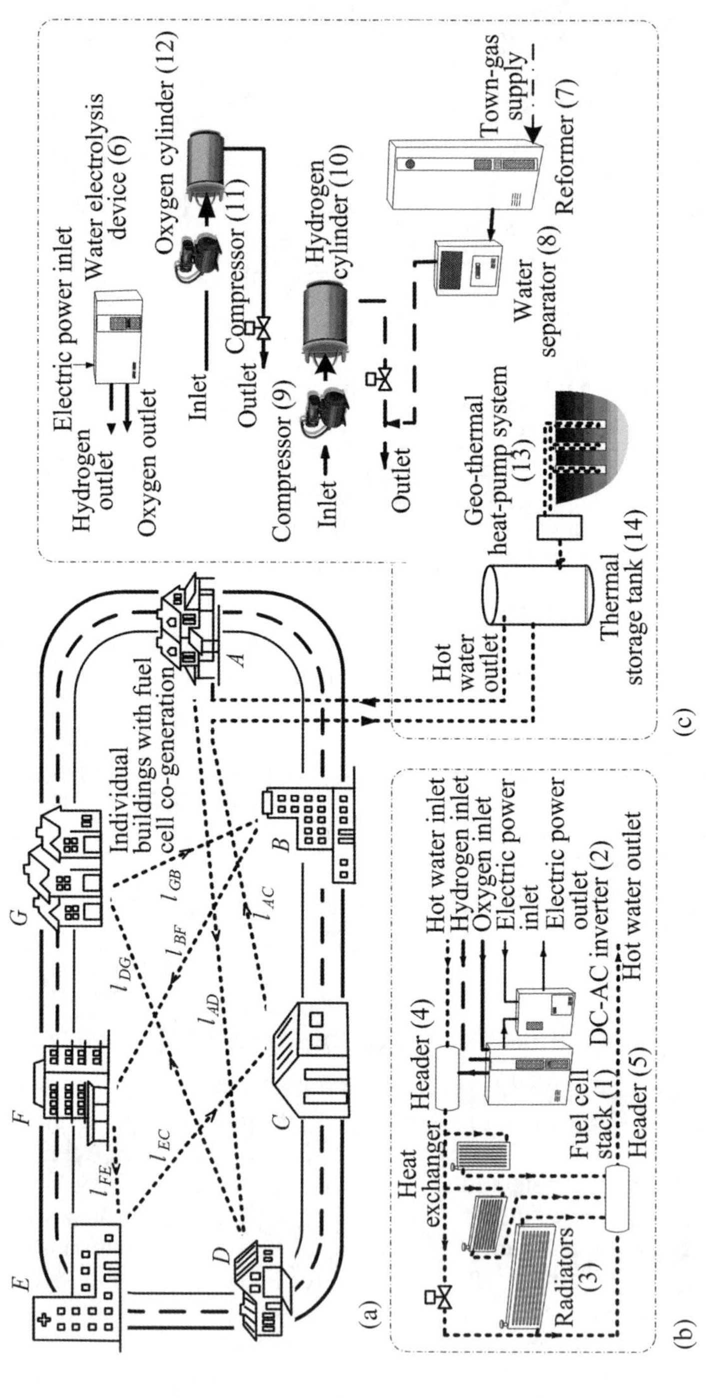

Figure 3.2 Fuel cell network system model (R2 type); (a) Fuel cell network model; (b) Fuel cell unit installed in building; (c) Auxiliary machine installed in utility room.

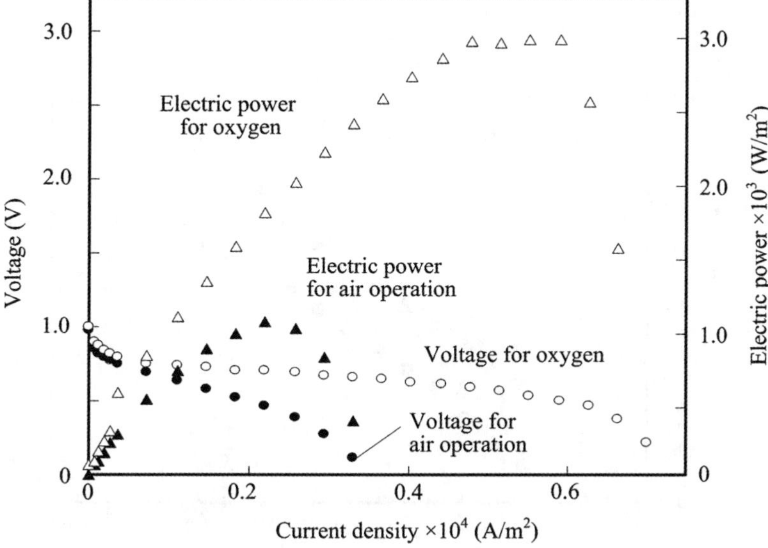

Figure 3.3 Cell performance.

R2 type and the building with a machinery room can be planned arbitrarily. In the example of Figure 3.2(a), hot water is supplied in the order of Building ADGBFECA. The machinery room of Figure 3.2(c) is installed in Building A. The equipment scheme installed in the building and machinery room in the R2 type is shown in Figures 3.2(b) and (c).

Ambient air is usually supplied to the fuel cell installed in R1 and R2 from a blower. However, both types can also supply oxygen through piping. Moreover, it is assumed that reformed gas of the city-gas reformer and hydrogen of the cylinder can be supplied to the fuel cells at arbitrary times through the network.

3.2.2 Power-Generation Characteristics of the Fuel Cell

Figure 3.3 shows the power-generation characteristics when hydrogen and oxygen are supplied, and when supplying hydrogen and air by the results of the performance measurement of a PEFC. The differences in these power-generation characteristics are considered to be due to the difference in oxygen partial pressure, the water balance inside the cell, and the electrical receptivity change of the ion-exchange membrane. The power-generation characteristics differ between supplying reformed gas to a fuel cell, and supplying hydrogen. However, since there are few differences in the power-generation characteristics of reformed gas or hydrogen, this difference is ignored.

Figure 3.4 shows the characteristics of the power and heat output when supplying air or oxygen to a cathode using the same fuel cell (the electrode

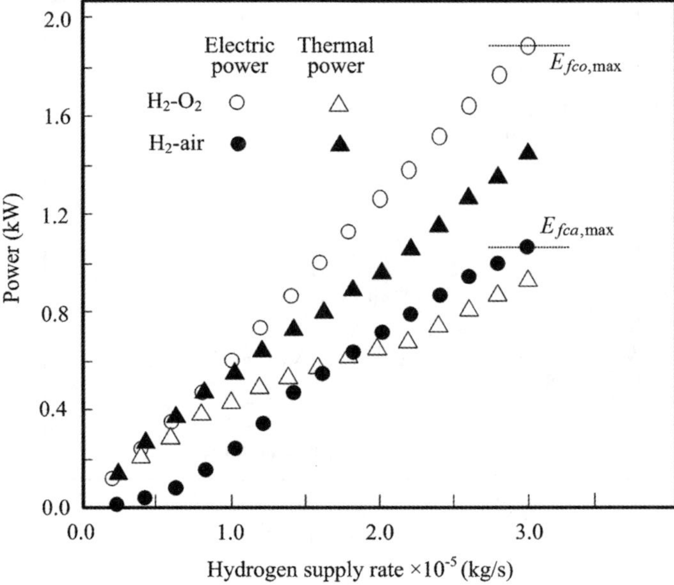

Figure 3.4 Output characteristics of a fuel cell.

surface is $1\,m^2$) as shown in Figure 3.3. The maximum power output when supplying air to the cathode is $E_{fca,\max} = 1.05\,kW$, and it is $E_{fco,\max} = 1.9\,kW$ in the supply of oxygen. In this way, if oxygen is supplied to the cathode, the power output will increase. Therefore, if oxygen is supplied and generated to a fuel cell when there is high power demand, the fuel cell can be miniaturized compared with the design capacity by air supply. If the fuel cell with the characteristics shown in Figure 3.4 is used with maximum output, the fuel cell facility capacity will decrease by the value of $(E_{fco,\max} - E_{fca,\max})$.

3.2.3 Load Leveling Using Water Electrolysis

Figure 3.5 shows the model indicating power demand amount $E_{need,t}$ to which is added the power demand amount of each building in Figure 3.1 or Figure 3.2 for every sampling time t. E_{sep} in this figure is the threshold value of the region of low load and high load. By using this threshold value, load leveling is attempted using the method described below. When $E_{need,t}$ is less than E_{sep}, it generates electricity by supplying reformed gas and air to the fuel cell. However, the production of electricity of the fuel cell is always E_{sep}, it supplies power whose value is the difference between E_{sep} and $E_{need,t}$ to the water electrolyzer, and produces hydrogen and oxygen (the black area in Figure 3.5). After compressing these gases, they are stored in each cylinder. When $E_{need,t}$ exceeds E_{sep}, it generates electricity by supplying the hydrogen and oxygen in the cylinders to the fuel cell through the network. In the proposed method of load

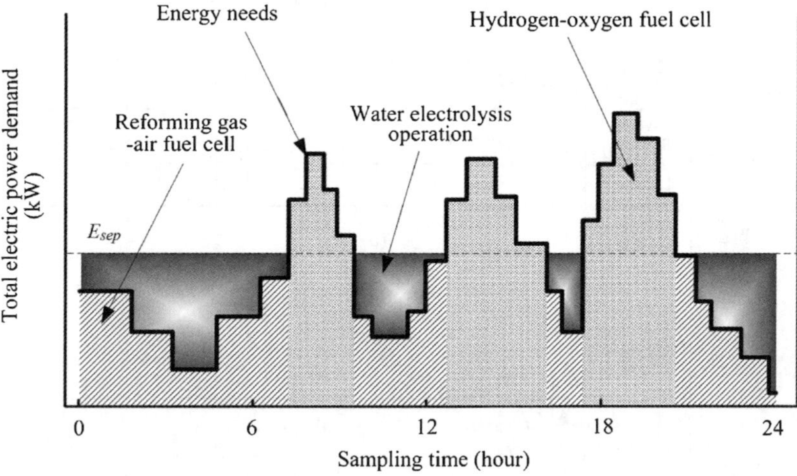

Figure 3.5 Fuel cell operation.

leveling, it is necessary to determine E_{sep}, where hydrogen and oxygen are produced at a time of low load, and the amount is consumed at a time of high load balance.

3.2.4 Distribution of the Fuel Cell

Figure 3.6 shows the model of (a) the hot-water piping route, (b) fuel cell capacity of each building, (c) change of hot-water temperature, and (d) piping heat release per unit length of the R1 type and R2 type. The machinery room of both types is installed in Building A, and hot water flows in the order of Building ABCDEFGA for the R1 type, and it flows in the order of ADGBFECA for the R2 type as shown in Figure 3.6(a). As shown in Figures 3.6(a) and (b), one fuel cell is installed in Building A ($F_{A'}$) for the R1 type, and the fuel cell of the capacity of F_A to F_G is installed in Building A to G for the R2 type. Hot water of temperature $T_{A,in,t}$ is input into Building A in the R1 type. Heat is supplied for the hot water from the fuel cell exhaust heat ($F_{A'}$), heat-storage tank, and geothermal heat pump, and as shown in Figure 3.6(c), hot water of temperature $T_{A,out,t}$ is output from Building A. After this, there is no heat input to for hot water, and hot water of temperature $T_{A,in,t}$ returns to the machinery room of Building A due to the heat consumption of Building B to G, and piping heat release. The temperature falls as the hot water of the R1 type progresses to Building G from Building A. Therefore, since the difference in temperature of the outside air and hot water is small, as shown in Figure 3.6(d), the piping heat release per unit length is small. On the other hand, in the R2 type, heat is supplied to hot water from distributed fuel cells. Therefore, the outlet hot-water temperature of each building fluctuates, as shown in Figure 3.6(c). As a

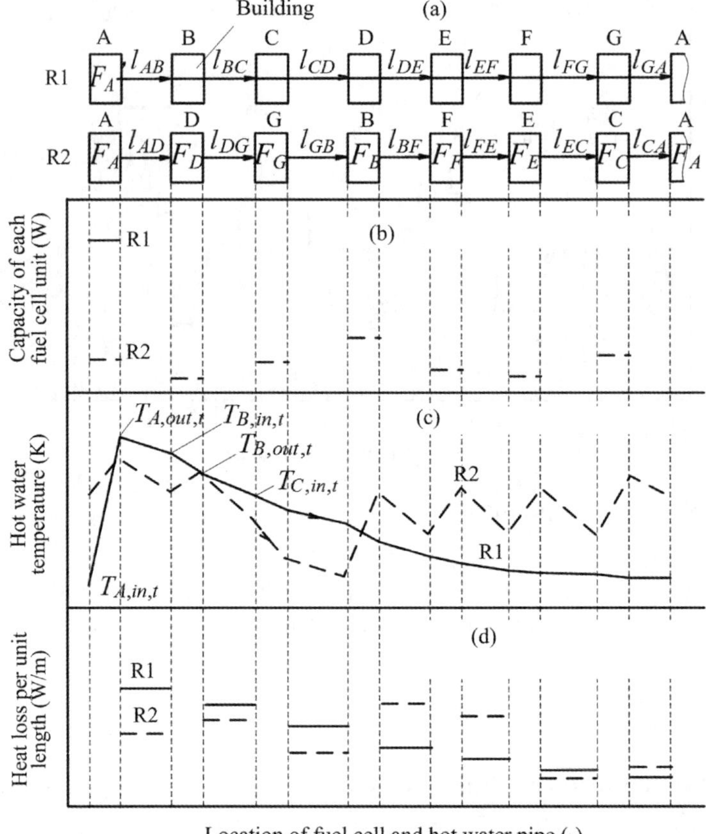

Figure 3.6 Arrangement plan of fuel cell units; (a) Hot water supply path; (b) Capacity of each fuel cell unit; (c) Hot-water temperature; (d) Heat loss of each hot-water piping.

result, the heat release per unit length of piping also fluctuates, as shown in Figure 3.6(d).

3.2.5 Energy-Balance Equation

At sampling time t, the water electrolyzer installed in the machinery room and the fuel cells of M installed in M buildings are operating ($M = 1$ in the R1 type). The power-balance equation (in this case is expressed with the following equation,

$$\sum_{m=1}^{M} E_{f,m,t} = \sum_{m=1}^{M} E_{need,m,t} + \Delta E_{el,t} + \Delta E_{hp,t} + \sum_{v=1}^{V} \Delta E_{sub,v,t} \tag{3.1}$$

The left-hand side of eqn (3.1) expresses the power output in the DC–AC converter outlet of the fuel cells of M. Moreover, the 1st term on the right-hand side is the power demand amount in each building, the 2nd term expresses the power consumption of the water electrolyzer, the 3rd term expresses the power consumption of the heat pump, and the 4th term expresses the power consumption of the auxiliary machinery (the pump of the hot-water network, and the compressor of hydrogen and oxygen).

The heat balance of the system is expressed below,

$$\sum_{m=1}^{M} H_{f,m,t} + H_{st,t} + H_{hp,t} = \sum_{m=1}^{M} H_{need,m,t} + \sum_{m=1}^{M} \Delta H_{hw,mm',t} \quad (3.2)$$

The 1st term on the left-hand side of eqn (3.2) expresses the exhaust heat of the fuel cell of M, and the 2nd term and the 3rd term express the heat output from the heat-storage tank and the heat pump, respectively. The right-hand side of eqn (3.2) expresses heat consumption, the 1st term is the heat demand of each building connected to the network, and the 2nd term expresses the heat release of the hot-water piping that connects each building. $\Delta H_{hw,mm',t}$ expresses the heat release of the hot-water piping that connects Building m to Building m', and is calculated from

$$\Delta H_{hw,mm',t} = h \cdot \pi \cdot D_p \cdot l_{mm'} \cdot (T_{m,out,t} - T_{atm,t}) \quad (3.3)$$

Equation (3.4) is the balance equation of hydrogen. The 1st term on the left-hand side of eqn (3.4) expresses the quantity of hydrogen production of the water electrolyzer, and the 2nd term expresses the hydrogen quantity supplied to the network from the cylinder, and the 3rd term expresses the quantity of hydrogen production of the reformer. Moreover, the right-hand side expresses the hydrogen consumption of the fuel cell of M. Equation (3.5) is a balance equation of oxygen. The 1st term on the left-hand side expresses the oxygen concentration of the water electrolyzer, the 2nd term expresses the amount of oxygen supplied from the cylinder, and the 3rd term expresses the amount of oxygen in the air supply of the blower. The right-hand side is the amount of oxygen consumed with the fuel cell.

$$Q_{el,H_2,t} + Q_{a,H_2,t} + Q_{r,H_2,t} = \sum_{m=1}^{M} Q_{f,m,H_2,t} \quad (3.4)$$

$$Q_{el,O_2,t} + Q_{a,O_2,t} + Q_{bw,O_2,t} = \sum_{m=1}^{M} Q_{f,m,O_2,t} \quad (3.5)$$

3.2.6 Operating Method of the System

The exhaust heat of each fuel cell connected to the network is used for buildings in which a fuel cell is installed, which is given priority. The surplus heat of each building is recovered in the hot-water network. On the other hand, when the heat of a certain building runs short, heat is received from the hot-water piping. Moreover, when the heat of the overall network system runs short, heat is supplied to the network from the heat-storage tank and the heat pump. When the network has excess heat, surplus heat is stored in the heat-storage tank. The heat pump is operated when the heat of the heat-storage tank is insufficient.

3.3 Analysis Method

3.3.1 Procedure of Analysis

The analysis follows steps (1) to (3).

(1) Load leveling using water electrolysis, and calculation of E_{sep}.

The load-leveling method using water electrolysis is employed in the R1 type (the R2 type also uses the same procedure). In order to determine E_{sep} in Figure 3.5, an initial value is decided at random concerning the given power demand pattern. At this time, the amount of production of hydrogen and oxygen in a low-load period is calculated, and hydrogen and oxygen consumption in a high-load period are also calculated. The balance is calculated from the amount of production and consumption of hydrogen and oxygen. The value of E_{sep} is changed, and it is repeatedly calculated until the balance of hydrogen and oxygen becomes sufficiently small. In the analysis case in Section 3.4, the time of less than 1% of the balance error was adopted. Balance eqn (3.1) of the power, balance eqn (3.4) of the hydrogen and eqn (3.5) of oxygen, are used for the calculation of the balance of hydrogen and oxygen. Figures 3.4 and 3.7 are used as the power-generation characteristics of the fuel cell and the characteristics of the water electrolyzer. When the fuel cell capacity in the analysis exceeds that of Figure 3.4, it is assumed that the relation of Figure 3.4 can be extrapolated.

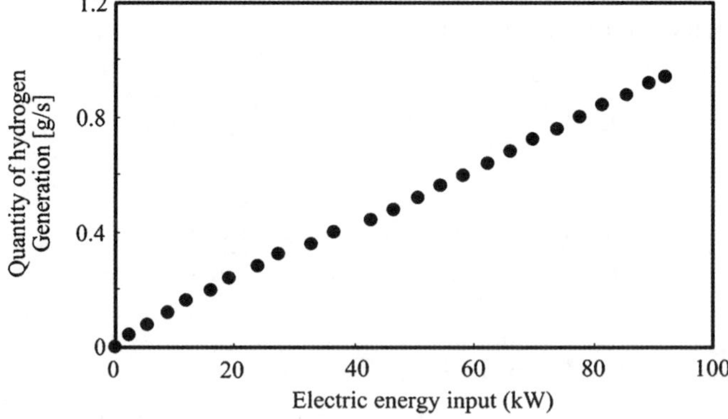

Figure 3.7 Characteristics of water electrolysis device.

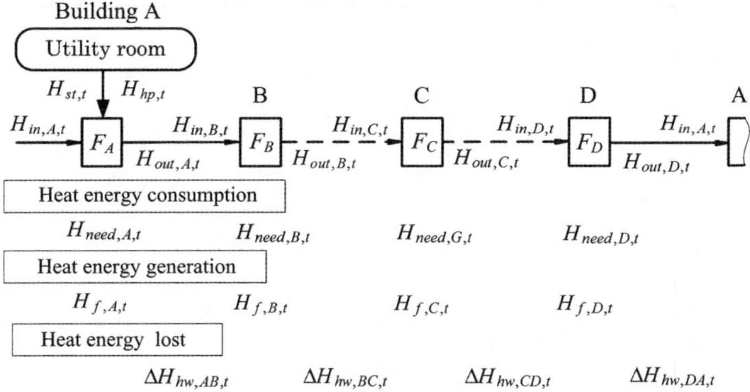

Figure 3.8 Heat energy network model.

(2) Calculation of heat release from the hot-water piping.

Figure 3.8 shows the heat release model of the hot-water piping. The fuel cell is installed in four houses, Buildings A to D. Each building is connected with piping, and hot water returns to Building A. The machinery room is set in Building A, and the heat outputs of heat-storage tank and heat pump installed in this machinery room are $H_{st,t}$ and $H_{hp,t}$. There is the heat demand of $H_{need,A,t}$ to $H_{need,D,t}$ in Building A to D, respectively. In the fuel cell installed in each building, there is exhaust heat power output of $H_{f,A,t}$ to $H_{f,D,t}$. Therefore, the heat balance of Buildings A to D is calculable from eqn (3.2). Moreover, the heat release (from $\Delta H_{hw,AB,t}$ to $\Delta H_{hw,DA,t}$) of the hot-water piping that connects each building is calculated using eqn (3.3). $T_{atm,t}$ in eqn (3.3) employs the outside air temperature in Tokyo as shown in Figure 3.9.[3]

(3) Route planning of hot-water piping considering heat-release loss.

Since a fuel cell is placed in each building for the R2 type, it is necessary to determine the capacity of each fuel cell. The outlet hot-water temperature of a certain building is decided by the heat balance in the building, and the heat release of the hot-water piping is calculated from the difference in the outlet hot-water temperature and the outside air temperature. Therefore, the heat release of the overall network differs according to the capacity of the fuel cell installed in each building. In this chapter, as shown in Figure 3.10, information on the capacity of the fuel cell installed in each building and the piping route is expressed with a gene, and these are installed into a genetic algorithm. It is evaluated as a solution with high fitness, so that there are few values in eqn (3.6) showing heat release from the hot-water piping. The calculation is iterated, chromosomes are evolved and a solution with high fitness is sought. In the last generation's chromosomes, a solution with the highest fitness is determined as an optimal solution. From the information on the optimum chromosome, the capacity of the fuel cell installed in each building and the piping route are determined,

$$F = \sum_{t=1}^{Period} \sum_{l=1}^{M} \Delta H_{hw,l,t} \tag{3.6}$$

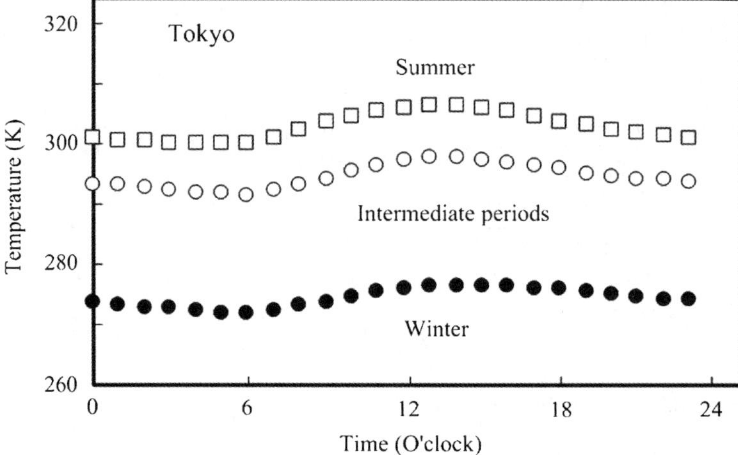

Figure 3.9 Outside air temperature.

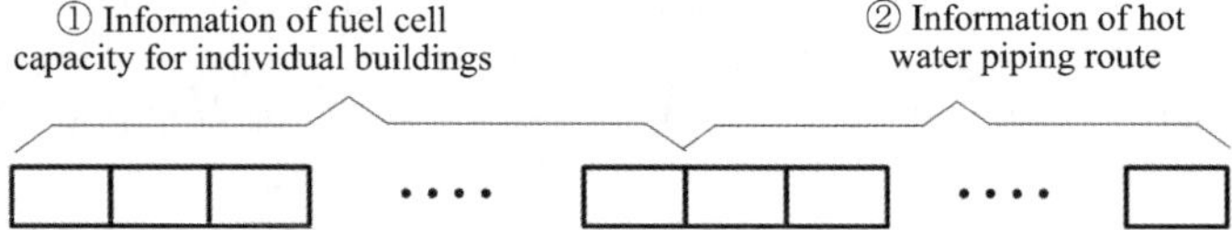

Figure 3.10 Chromosome model.

3.3.2 Solution Parameters

As parameters of the genetic algorithm employed in the analysis in Section 3.4, the population is 10 000, the generation number is 20, and the crossover probability is 0.5. The gene manipulation of mutation is not added. Search of the hot-water piping route in the R1 type is also analyzed using the genetic algorithm.

3.4 Case Study

3.4.1 Energy Demand Pattern and Network System

In this case study, an energy network composed of seven buildings is investigated. The energy-need pattern in winter (January), mid-term (May), and summer (August) of each building is shown in Figure 3.11.[4–6] These energy-demand patterns are assumed to be in Tokyo. Space-cooling power in summer is included in the power demand shown in Figure 3.11, and hot-water supply and space heating are included in the heat demand. However, the heat for

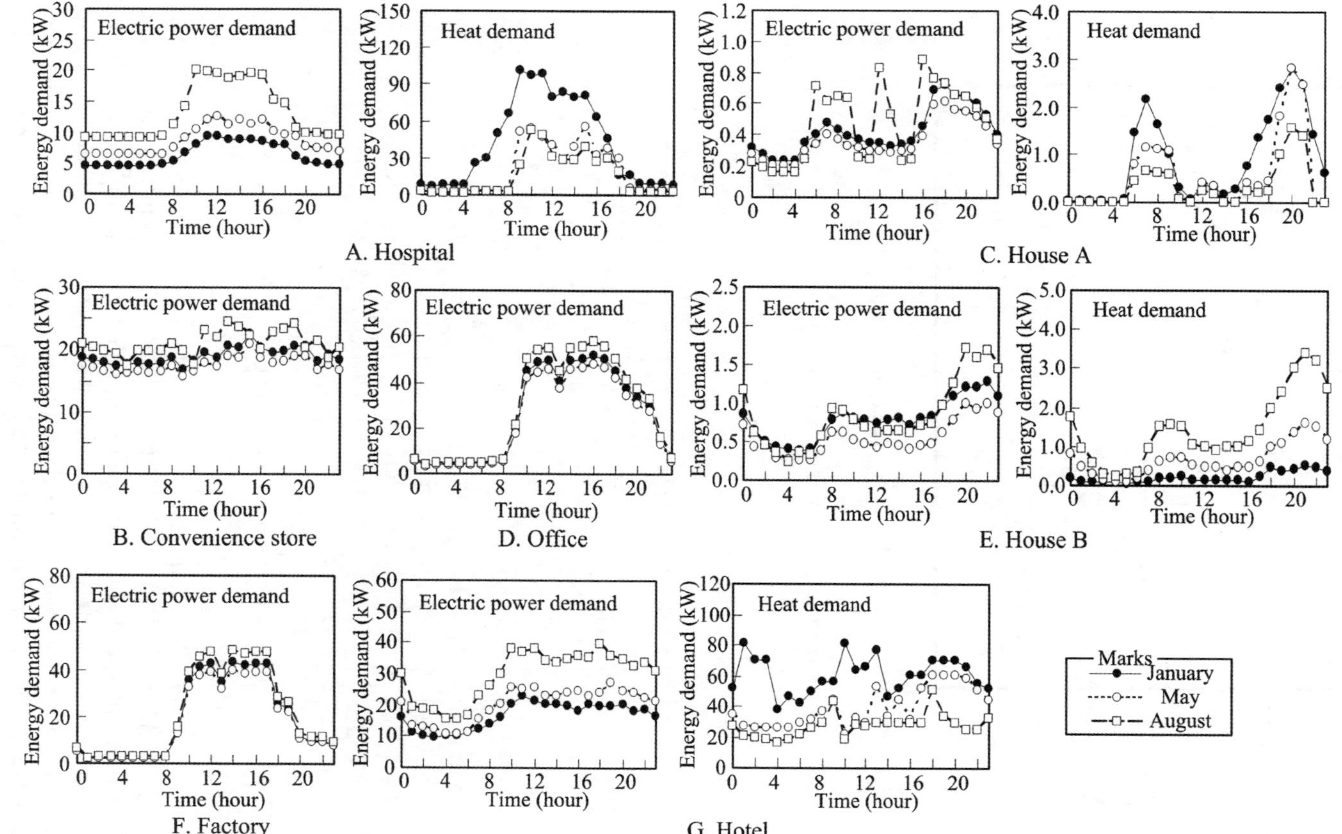

Figure 3.11 Energy-demand patterns.

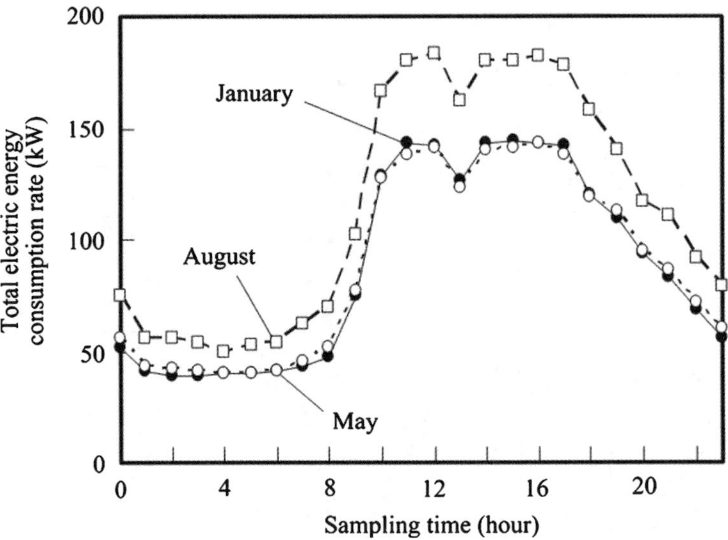

Figure 3.12 Demand patterns of total electric energy.

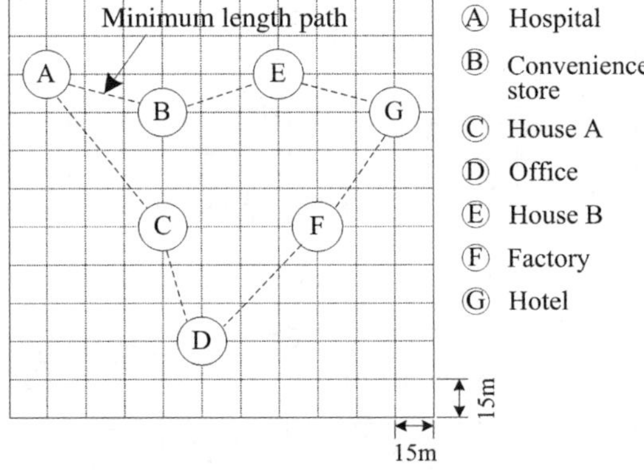

Figure 3.13 Location model of hot-water piping.

convenience stores, offices, and factories is supplied from an electric heat pump. Figure 3.12 shows the sum of the power demand amount of these seven buildings. The arrangement of the buildings is shown in Figure 3.13. Moreover, the broken line in Figure 3.13 is the hot-water piping route of the shortest distance. In these analyses, in order to make the hot-water flow rate in the piping 1 m/s or less, the inside diameter of the piping was set at 60 mm. The hot-water piping is equipped with a 40-mm-thick polystyrene-foam system heat insulating mould. Moreover, the overall heat-transfer coefficient on the surface of the heat insulating mould is set at 8.0 W/m^2 K. Under the conditions described above, reduction in fuel cell capacity using load leveling, and the route of the hot-water piping for minimum heat release are investigated.

3.4.2 Reduction Effect of Fuel Cell Facility Capacity

When the threshold value E_{sep} of a low-load region and a high-load region is calculated according to the procedure of Section 3.3.1 (1), a representative day in January is 109 kW and a representative day in May and August is 125 kW. By installing E_{sep} into the load leveling described in Section 3.2.3, the fuel cell is made to follow the power load pattern of Figure 3.12. Figure 3.14 shows the fuel cell exhaust heat in this case. In the heat balance on a representative day in January shown in Figure 3.14, since heat runs short in the 7:00 to 17:00 period, the heat pump is operated. On the other hand, the heat supply and demand on representative days in May and August show much heat surplus. Moreover, Figure 3.15 shows the calculation result of the electrode surface of the fuel cell at the time of installing E_{sep} and performing load leveling. The fuel cell electrode surface in Figure 3.15 expresses fuel cell capacity. The data enclosed within the broken line in Figure 3.15 are power generation using hydrogen and oxygen. These gases are produced using the power generated with reformed gas and air. The fuel cell generated with reformed gas and air is operated at a time other than the broken-line region in Figure 3.15. At the time of high load from 12:00 to 16:00 on a representative day in August, an ~180-m^2 electrode surface

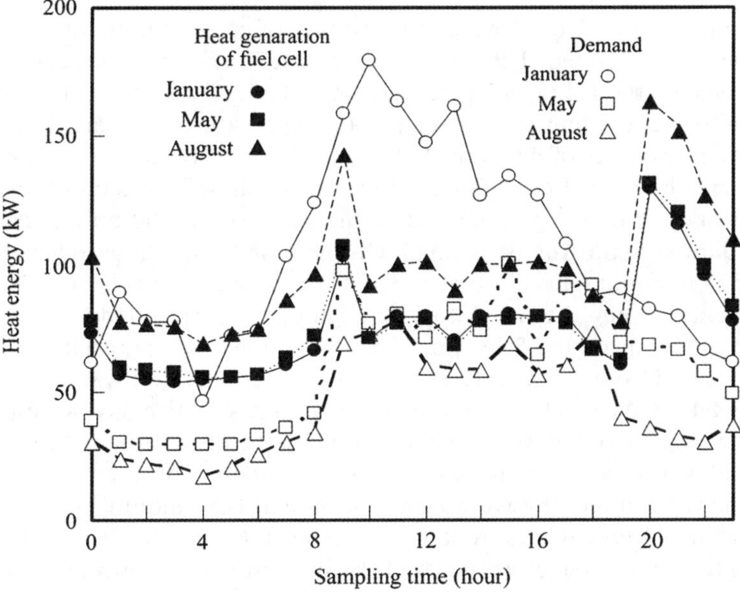

Figure 3.14 Demand patterns of total heat energy.

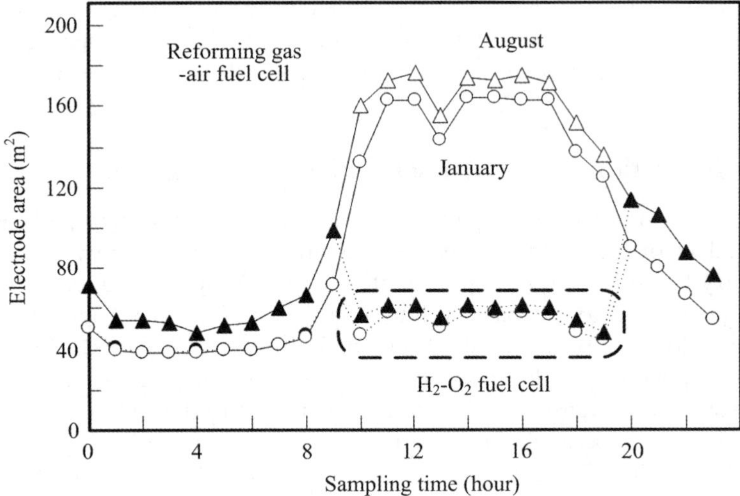

Figure 3.15 Results of electrode area.

was conventionally taken. If load leveling using water electrolysis is employed, the fuel cell can be reduced to a 120-m^2 electrode surface, which is equivalent to 2/3 at the peak at 20:00.

3.4.3 Route Planning Result of Hot-Water Piping

The result of the outlet hot-water temperature of each building that composes the network is described. The outlet hot-water temperature differs according to the R1 type or the R2 type. Moreover, since the heat release of the hot-water piping differs according to the outside air temperature, the sampling time is different. The results of 4:00 and 16:00 on representative days in January and August are shown in Figure 3.6. As Figure 3.12 shows, the sum of the power demand of each building connected to the network at 4:00 on representative days in January and August is small. On the other hand, this value is large at 16:00. The horizontal axis in Figure 3.16 is the route order (No. 1 to No. 7) of the hot-water piping. Letters A to G in Figure 3.16 correspond to the building number shown in Figure 3.13. For example, in the analysis result at 4:00 and 16:00 for the R1 type on a representative day in January, hot water flows in the order of GFDCABE. The optimal path on a representative day in January for the R1 type is GFDCABE, and the optimal path on a representative day in August is BEGFDCA. In this way, the starting points of the hot water differ according to each month. Moreover, the route of both month's representative days for the R2 type is GEBACDF. All routes GFDCABE, BEGFDCA, and GEBACDF are the same as a result of the shortest route shown in Figure 3.13. However, GFDCABE and BEGFDCA are clockwise rotations and GEBACDF is counterclockwise. The outlet hot-water temperature of each

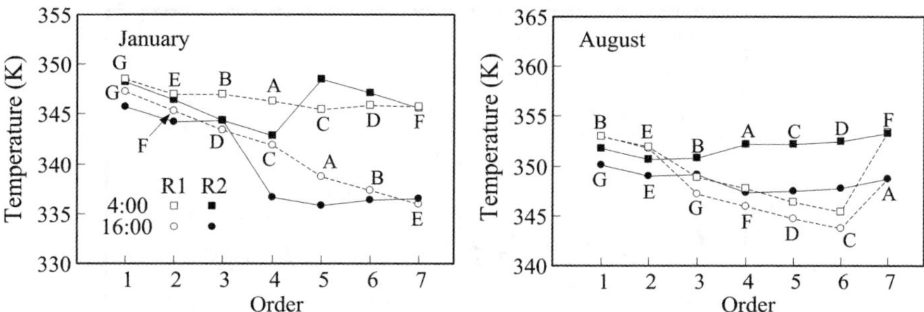

Figure 3.16 Hot-water temperature of outlet piping of buildings in January and August.

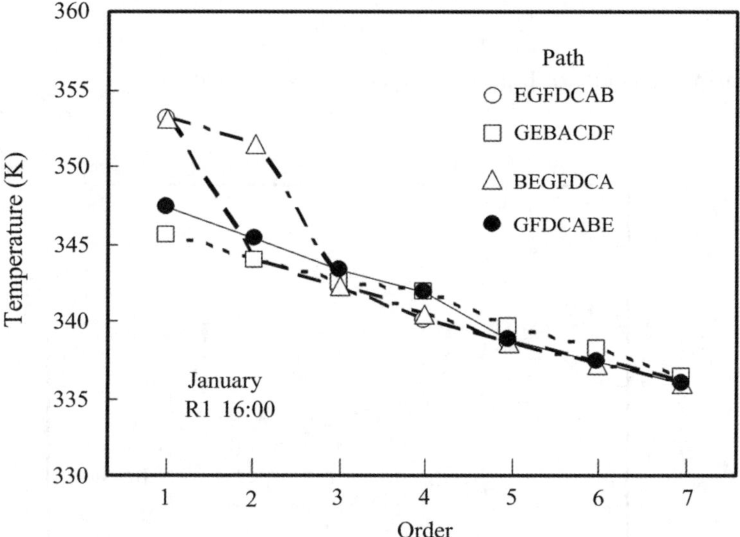

Figure 3.17 Hot-water temperature of outlet piping of buildings.

building differs in the starting point of the hot-water piping, route, and flow direction, as shown in Figure 3.17. This shows the result of the hot-water temperature when setting the starting point of the hot-water piping as B, E, or G. Figure 3.18 shows the result of the hot-water piping heat release relevant to the piping routes in Figure 3.17. Figure 3.19 shows the result of the piping heat release in the network on a representative day every month. Under these analysis conditions, the difference in the heat release for the R1 type and R2 type on a representative day is less than 3% every month. Considering the analysis error of the genetic algorithm, these can be estimated as the same value. Therefore, if the heat release of the R1 type and R2 type is optimized, it will converge on almost the same value. However, since the R1 type in this case

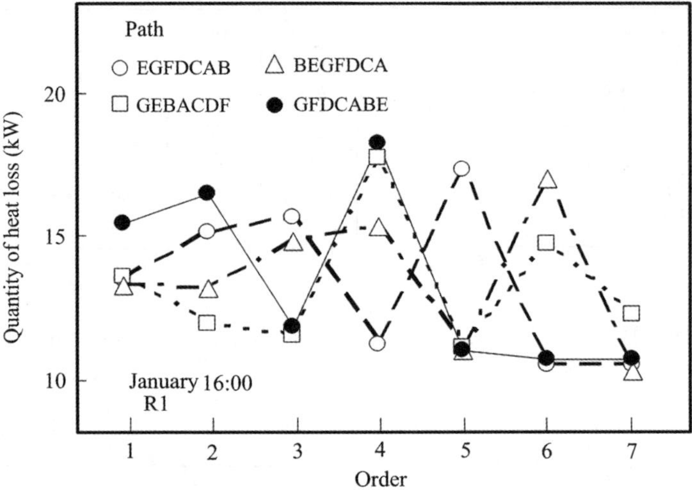

Figure 3.18 Quantity of heat loss of piping between buildings.

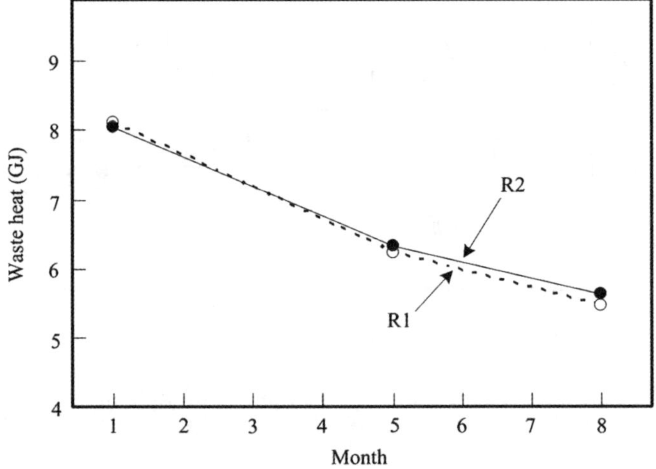

Figure 3.19 Waste heat of hot-water supply.

assumes that the starting point of the hot-water piping is movable to arbitrary buildings according to the month, it is not realistic.

3.4.4 Result of a Fuel Cell Arrangement Plan

Figure 3.20 shows the result of the fuel cell arrangement plan for the R2 type. The fuel cell capacity installed in each building is a circle of the broken line in

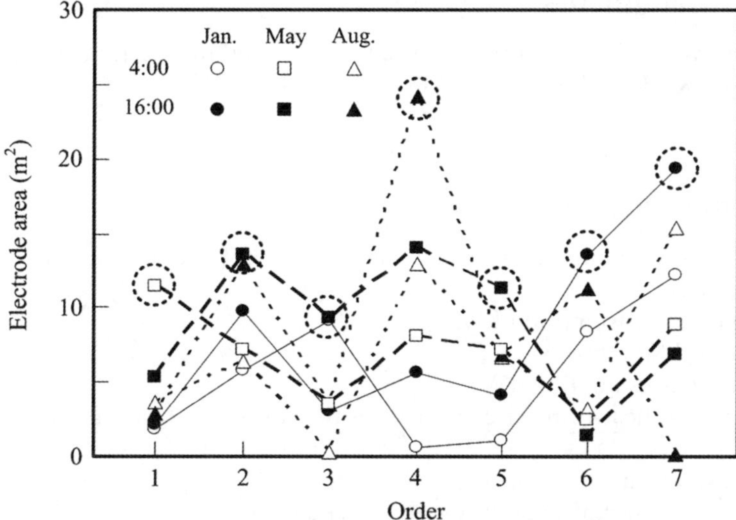

Figure 3.20 Installing capacity of fuel cell.

Figure 3.20. When the electrode surface of each building shown in Figure 3.20 is added, it is 97 m^2. The electrode surface when installing the installed capacity reduction by load leveling is 120 m^2. Furthermore, if the optimum arrangement plan of a fuel cell is installed, the electrode surface will be reduced to 97 m^2. When load leveling using water electrolysis and optimization of fuel cell distribution are installed, the fuel cell electrode surface is reduced by 46% compared to the conventional system.

3.5 Conclusions

For the fuel cell energy network system, the load-leveling method that supplies air and water electrolysis oxygen to the cathode of the fuel cell was proposed. Furthermore, the optimum operation plan of the hot-water network was proposed, and the fuel cell capacity of each building, position of the machinery room, piping route, and hot-water flow direction were investigated. The fuel cell energy network composed of individual houses, a hospital, a hotel, a convenience store, an office building, and a factory was analyzed, and the following conclusions were obtained.

(1) If the load-leveling method is used, the installed capacity of the fuel cell will be reduced by 34% compared with the conventional system.

(2) Moreover, when the fuel cell distribution is optimized, in accordance with the effectiveness of (1), there is a 46% reduction compared with the conventional system.

Acknowledgments

This work was partially supported by a Grant-in-Aid for Scientific Research(C) from JSPS.KAKENHI (17510078).

Nomenclature

D_p: heat-insulating-mold outside diameter of hot-water piping m
E: power kW
ΔE: power consumption kW
H: heat kW
ΔH: heat consumption kW
h: overall heat-transfer coefficient in the surface of heat insulating mould $W/m^2 K$
l: distance m
M: the number of buildings with a fuel cell
T: temperature K
t: sampling time
V: the number of auxiliary machinery with electricity consumption

Subscripts

a: cylinder
atm: open air
bw: blower
el: electrolyzer
f: fuel cell
fca: fuel cell generated with air
fco: fuel cell generated with oxygen
hp: heat pump
hw: heat release of hot-water piping
m: the number of each building
need: energy needs
r: reformer
sep: threshold value of low loading and high load
st: heat-storage tank
sub: auxiliary machinery

References

1. S. Obara and K. Kudo, Study on Improvement in Efficiency of Partial Load Driving of Installing Fuel Cell Network with Water Electrolysis Operation, *Trans. Jpn. Soc. Mech. Eng., Series B*, 2005, **71**(701), 237–244 (in Japanese).

2. M. Badami and C. Caldera, Dynamic Model of a Load-Following Fuel Cell Vehicle: Impact of the Air System, *SAE Technical Paper*, 2002, SAE-2002-01-100, 1–10.
3. U. Inoue, *Air-conditioning handbook*, 1996, 14, Maruzen (in Japanese).
4. Architectural Institute of Japan, The nationwide research study concerning the energy consumption in the house in the 2001 fiscal year, 2002, **3**, 3–6 (in Japanese).
5. Y. Yamano, Development of a load-levelling technique, *Denki*, 2002, **629**, 56–61 (In Japanese).
6. S. Ozaki and I. Tuziki, Trial calculation of the quantity of public electric power and city gas to be replaced by a distributed energy system, In: *Proceedings of the 9th Energy-resources Seminar*, 1990, **9**, 174–179 (in Japanese).

CHAPTER 4

Power-Independent House Using PEFC – Operation Plan of a Combined Fuel Cell Cogeneration, Solar Module, and Geothermal Heat Pump System

SHIN'YA OBARA

4.1 Introduction

From deregulation of the energy business, and an environmental problem, the installation spread of the small-scale distribution power due to a fuel cell and a heat engine is expected. Under the objective function set up by the designer or the user, optimization planning that controls small-scale distribution power is required. In dynamic operation planning of the energy plant, the analysis method using mixed integer linear programming has been developed.[1,2]

For the compound energy systems of solar modules and fuel cell cogeneration, there have been no reports of the optimization of operation planning. Therefore, there are no results showing the relationship between the objective function given to the combined system and operation planning. For examples such as solar modules or wind power, green-energy equipment is accompanied by the fluctuation of output in many cases. Almost all green energy equipment

RSC Energy Series No. 3
Compound Energy Systems: Optimal Operation Methods
By Shin'ya Obara and Arif Hepbasli

Published by the Royal Society of Chemistry, www.rsc.org

requires backup by commercial power, fuel cells, heat engines, *etc.* Operation planning of the system that utilizes renewable energy differs in the objective function and power and heat load pattern. Thus, this chapter investigates the operation planning of the compound energy system composed of proton-exchange membrane fuel cell (PEFC) cogeneration with methanol steam-reforming equipment, a solar module, geothermal heat pump, heat storage, water-electrolysis equipment, commercial power, and a kerosene boiler. In such a complex energy system, facility cost is high. However, in this chapter, we investigate an example of an independent power source with renewable energy. This chapter considers the operation planning of a system, and the optimization of equipment capacity. The genetic algorithm (hereafter described as GA) applicable to a nonlinear problem with many variables is installed into the optimization calculation of the operation planning of the system.[3] In the operation analysis of a complex energy system, mixed integer programming (MIP) other than GA can be used. Because the nonlinear analysis using MIP is made to approximate using a linear expression of relations, it is considered that the error is large. On the other hand, a GA is applicable to the analysis of the nonlinear problem of many variables. It is known that the analytic accuracy of the optimization using a GA can be used industrially. In a GA, the design variable of energy equipment is shown with many gene models. In this chapter, the objective functions given to the system were set up as (1) Minimization of error in demand-and-supply balance, (2) Minimization of the operation cost (fuel consumption) of energy equipment, (3) Minimization of the carbon dioxide gas emission accompanying operation, and (4) Minimization of the three objective functions described above. The load pattern in winter (February) and summer (August) of an average individual house in Sapporo, Japan, is used for the energy-demand model shown with a case study.[4] This chapter describes the operation plan of the independence energy system when installing a methanol steam-reforming type fuel cell and renewable energy into a cold-region house. Such complex operation optimization of the energy system has not been reported until now. Consequently, the method of installing and analyzing the GA application to the nonlinear problem of many variables was proposed. In terms of equipment cost, it is difficult for a proposed system to spread generally. However, the installation to the area where the commercial power is not fixed is possible.

4.2 Fuel Cell, Solar Modules, and Geothermal Heat Pump Combined System

4.2.1 Scheme of Combined System

Figure 4.1 shows the energy system scheme examined in this chapter. A combined system consists of a solar module (18), PEFC-CGS (the fuel cell) is (1), the reforming equipment is (2)–(5) and (12), geothermal heat pump, (17), boiler, (8), commercial electric power, heat-storage tank (10), and the water-electrolysis

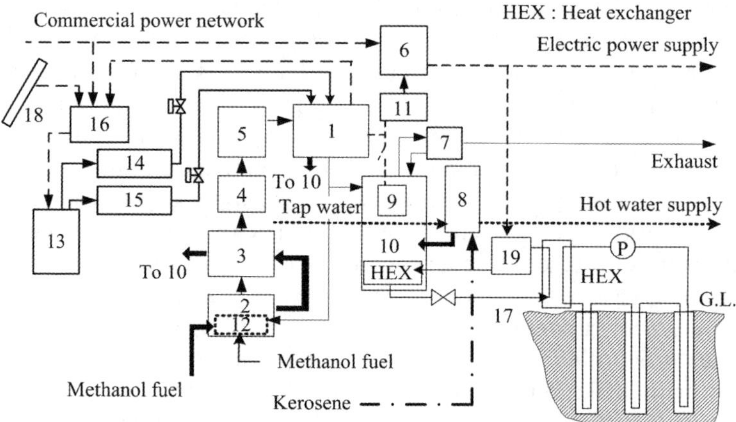

Figure 4.1 PEFC-CGS, heat pump and solar module combined system for houses.

equipment is (13)–(15). Water-electrolysis equipment is used to store electrical power with hydrogen and oxygen. The arrowhead in this figure shows the substance or direction of energy flux. Each system of the solar module, commercial electrical power, and PEFC-CGS is changed with a changeover switch (6), and electrical power is supplied to the consumer. However, electrical power is not supplied at once to the demand side from two or more power systems.

An electric heater (9) is installed inside the heat-storage tank, and electric power is changed into heat and can be stored. Hydrogen and oxygen can be produced if electric power is supplied to an electrolysis tank (13). Hydrogen and oxygen are stored in tanks (14) and (15), respectively, and these are supplied to PEFC and can be generated at any time. When the heat produced by the geothermal heat pump exceeds the quantity demanded, surplus heat is stored in the heat-storage tank. Although the exhaust heat of PEFC and the methanol steam-reforming equipment is also supplied to the heat-storage tank, when the total amount of heat exceeds the heat storage capacity, heat is radiated with a radiator (7). Tap water has heat exchanged for the heat-transfer medium inside the heat-storage tank, and moreover controls the temperature of this tap water by the boiler, and supplies hot water to the consumer.

Methanol fuel is supplied to the reformed gas system of the methanol steam-reforming equipment, and the catalytic-combustion equipment (12) installed in the evaporator (2). Kerosene fuel is supplied to the boiler (8). The energy-demand pattern used for analysis is a model in February (winter) and August (summer) in an average individual house in Sapporo in Japan, and shows this in Figure 4.2.

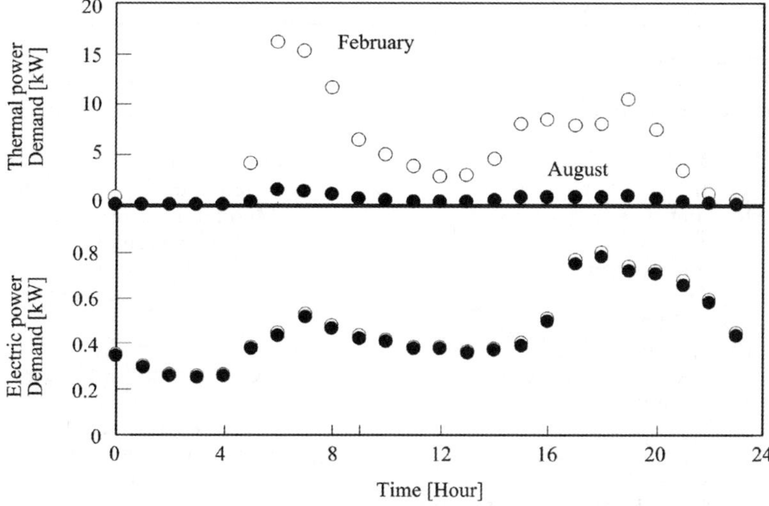

Figure 4.2 Energy demand of Sapporo city.

For Sapporo, a cold, snowy area, the annual average temperature is 288 K, and the mean temperature in February and August is 269 K and 294 K, respectively. The operating period of a system is made into 23:00 from 0:00 of a representative day, and sampling time is expressed by t_k ($k = 0, 1, 2,...,23$). The initial values of the capacity of each energy device set up the value used for the usual individual house. The specifications of each energy device are shown in Table 4.1.

Compared with the condition of the steady operation of the methanol reformer, the characteristics of a startup and a shutdown differ greatly. Cold-start operation and shutdown operation require about 20 min, respectively. In the analysis of this chapter, it is assumed that the startup of the methanol reformer is always a hot start.

4.2.2 Relational Expression

4.2.2.1 Energy Output of PEFC-CGS

A 33 kW methanol steam-reforming type PEFC shown in Figure 4.3 is used for the output characteristic of the fuel cell introduced into analysis.[5] The horizontal axis of Figure 4.3 is divided into two or more zones, and the output characteristics are given by the analysis program by using the secondary least-squares method approximation for each range. The electric power output at the time of supplying and generating hydrogen and oxygen stored by water electrolysis to a fuel cell is expressed by

$$E_{FS,t_k} = I_{t_k} \cdot E_{V,t_k} - \Delta W_{FS,t_k} = \frac{Q_{f,t_k} \cdot F_d}{E_c} \cdot E_{V,t_k} - \Delta W_{FS,t_k} \tag{4.1}$$

Table 4.1 Energy device specifications.

Solar module	
Area	6.0 m^2
Electric energy output	3 kW (Maximum)
Fuel cell	
Type	Proton-exchange membrane fuel cell
Fuel	Water/methanol = 1.4/1.0 (mole ratio)
Reforming type	Methanol steam reforming
Electric energy output	Maximum 3 kW
Thermal energy output	9 kW (Maximum)
	5 kW (Maximum)
Commercial power	
Heat pump	
Type	Geothermal heat source
Energy source	Electricity
p–h diagram	See Figure 4.4
Thermal energy output	5 kW (Maximum
COP	3.0
Electrolysis device [8]	
Electrolysis efficient	0.85 (Constant)
Accumulation of electricity	180 MJ
Backed boiler	
Fuel	Kerosene
Efficiency	0.85 (Constant)
Thermal energy output	40 kW (Maximum)
Thermal storage tank	
Thermal storage capacity	180 MJ
Heat medium temperature	353 K (Maximum)
Thermal storage efficiency	0.95

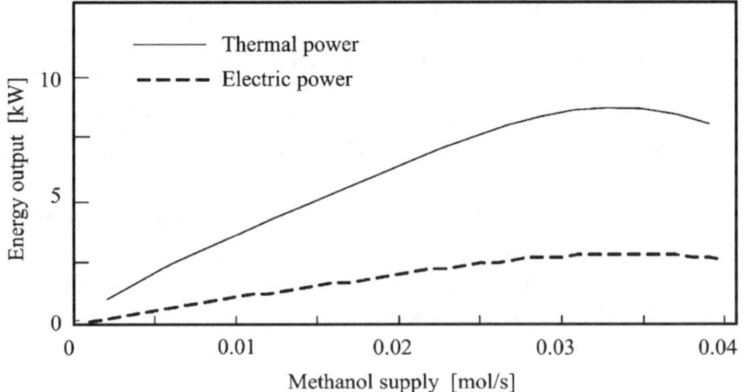

Figure 4.3 Characteristics of fuel cell stack with methanol steam reforming.

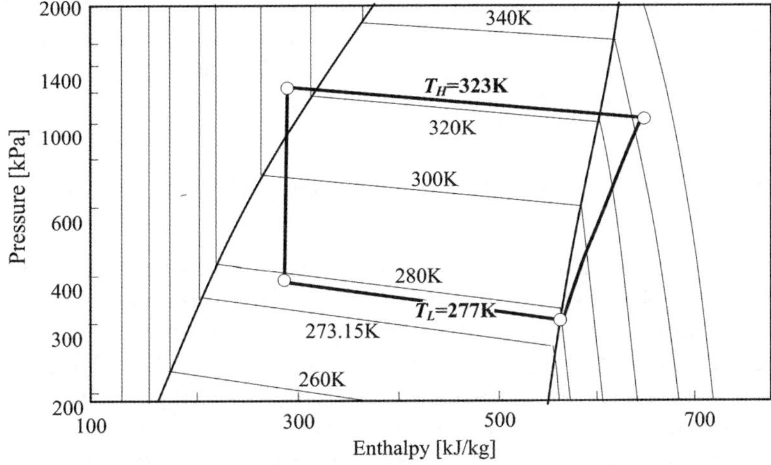

Figure 4.4 *p–h* diagram of Refrigerant AHC-12a (HC-TECH Inc., 1997).

Here, I_{t_k}, E_{V,t_k}, $\Delta W_{FS,t_k}$, Q_{f,t_k}, F_d and E_c express current, voltage, power loss of a cell stack, hydrogen supply, Faraday constant and chemical equivalent, respectively.

4.2.2.2 Heat Output of Geothermal Heat Pump

Figure 4.4 is a *p–h* diagram of Refrigerant HC-12a (OZ Technology, U.S., State of Idaho), used by the geothermal heat pump.[6,7] This refrigerant is a mixed refrigerant of propane, butane, and isobutene. Although the output characteristics of the heat pump were the analysis of soil temperature T_L and condensation temperature T_H exactly, the coefficient of performance COP_{t_k} was set to 3.0 in this chapter.

4.2.2.3 Characteristic Equation of Water-Electrolysis Tank

From sampling time t_k to Δt, the electric power supplied to an electrolysis tank is expressed by E_{EL,t_k}, and the efficiency of the electrolysis tank is expressed by ϕ_{EL}. In this case, hydrogen quantity $Q_{H_2 t_k}$ to be produced is calculated by eqn (4.2). Moreover, the amount of oxygen produced is similarly calculated. The efficiency of the water electrolyzer refers to the results of a study[8] that used the solid polymer membrane, and ϕ_{EL} is set as 0.85.

$$Q_{H_2,t_k} = \frac{E_{EL,t_k} \cdot E_c}{F_d \cdot E_V} \cdot \phi_{EL} \tag{4.2}$$

In calculations for this case study, the hydrogen and oxygen are pressurized to 1.0 MPa, respectively. The work of the compressor is assumed to be compression work for an ideal gas. The whole compressor efficiency including an inverter controller loss and the power consumption in an electric motor, transfer loss of power, loss with insufficient air leak and cooling, and other machine losses is set up to 50%.

4.2.2.4 Characteristic Equation of Heat-Storage Tank and Boiler

The conditional expression showing heat-storage characteristics is given by eqns (4.3) and (4.4) using the amount of maximum heat storage $S_{St,\max}$, and maximum temperature $T_{St,\max}$ of the heat medium. The capacity and the specific heat of the heat-storage medium are expressed by V and C_p, and the outside air temperature is expressed by $T\infty$ (The heat medium is assumed to be calcium chloride.) Moreover, the heat-storage temperature at time t_k is calculated by $T_{St,t_k} = S_{St,t_k}/(\rho \cdot C_p \cdot V)$. Here, ρ express the density of the heat-storage medium.

$$0 \leq S_{St,t_k} \leq S_{St,\max} \tag{4.3}$$

$$T_{\infty,t_k} \leq T_{St,t_k} \leq T_{St,\max} \tag{4.4}$$

The characteristic equation (of heat-storage tank between time t_k and Δt is given by

$$S_{St,t_k} - S_{St,t_{k-1}} = \left\{ H_{St,in,t_k} - H_{St,out,t_k} - \phi_{St} \cdot \rho \cdot C_p \cdot V \cdot \left(T_{St,t_k} - T_{\infty,t_k} \right) \right\} \cdot \Delta t \tag{4.5}$$

H_{St,in,t_k} and H_{St,out,t_k} express the heat input and heat output of the heat-storage tank, respectively. The third term in the right-hand bracket of eqn (4.5) includes outside air temperature T_{∞,t_k}, assuming heat storage loss is dependent on outside air temperature. However, in the analysis in this chapter, the efficiency of heat storage ϕ_{St} is set to 0.95, and change in outside air temperature is not taken into consideration. Figure 4.5 shows the relationship between the fuel consumption of a boiler and hot-water-supply output. It is expressed with the calorific value of the fuel being α_{Boiler}, the boiler efficiency being ϕ_{Boiler}, and the

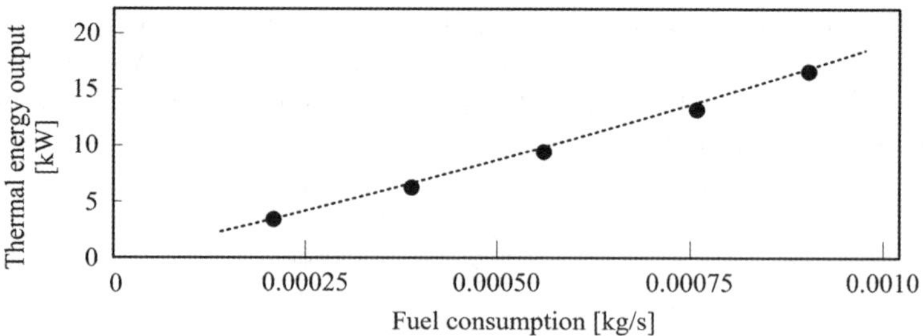

Figure 4.5 Thermal energy output of boiler.

fuel-supply flow being F_{Boiler,t_k}, and the characteristic equation of a boiler is given by the following equation,

$$H_{Boiler,t_k} = \alpha_{Boiler} \cdot F_{Boiler,t_k} \cdot \phi_{Boiler} \tag{4.6}$$

4.2.2.5 Characteristic Equation of Solar Module

Figure 4.6 shows the results of measurement of the production of electricity of the solar module in February and August in Sapporo.[9] However, the panel was vertically installed so that this solar module would not be covered in snow in winter. Therefore, the production of electricity decreases as shown in the results of Figure 4.6 at 13:00.

4.2.3 Energy Supply Path

The energy equipment is expressed by D_i, and let subscript i ($i = 1,2,3,\ldots,M$, M are the number of pieces of equipment) be the equipment number. The electric power and heat that are output by energy device D_i follow one path (a) to (j) as shown in Figure 4.7. When electric energy E_{D_i,t_k} generated by the system exceeds power demand E_{Need,t_k}, hydrogen and oxygen are produced and stored by water electrolysis. Moreover, it is also possible to change electric power into heat with a heater, to shift time, and to supply the demand side.

4.3 Energy Balance and Objective Function

4.3.1 Objective Function of System

The objective function given to the system is given by (1) Minimization of error in demand-and-supply balance, (2) Minimization of the operation cost (fuel

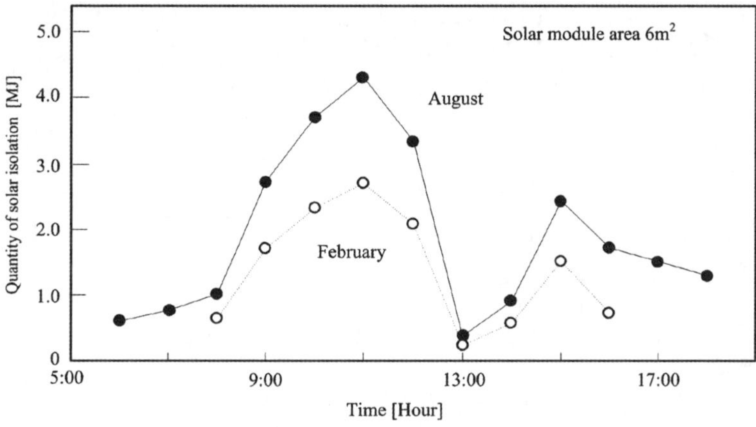

Figure 4.6 Time change of solar module output.

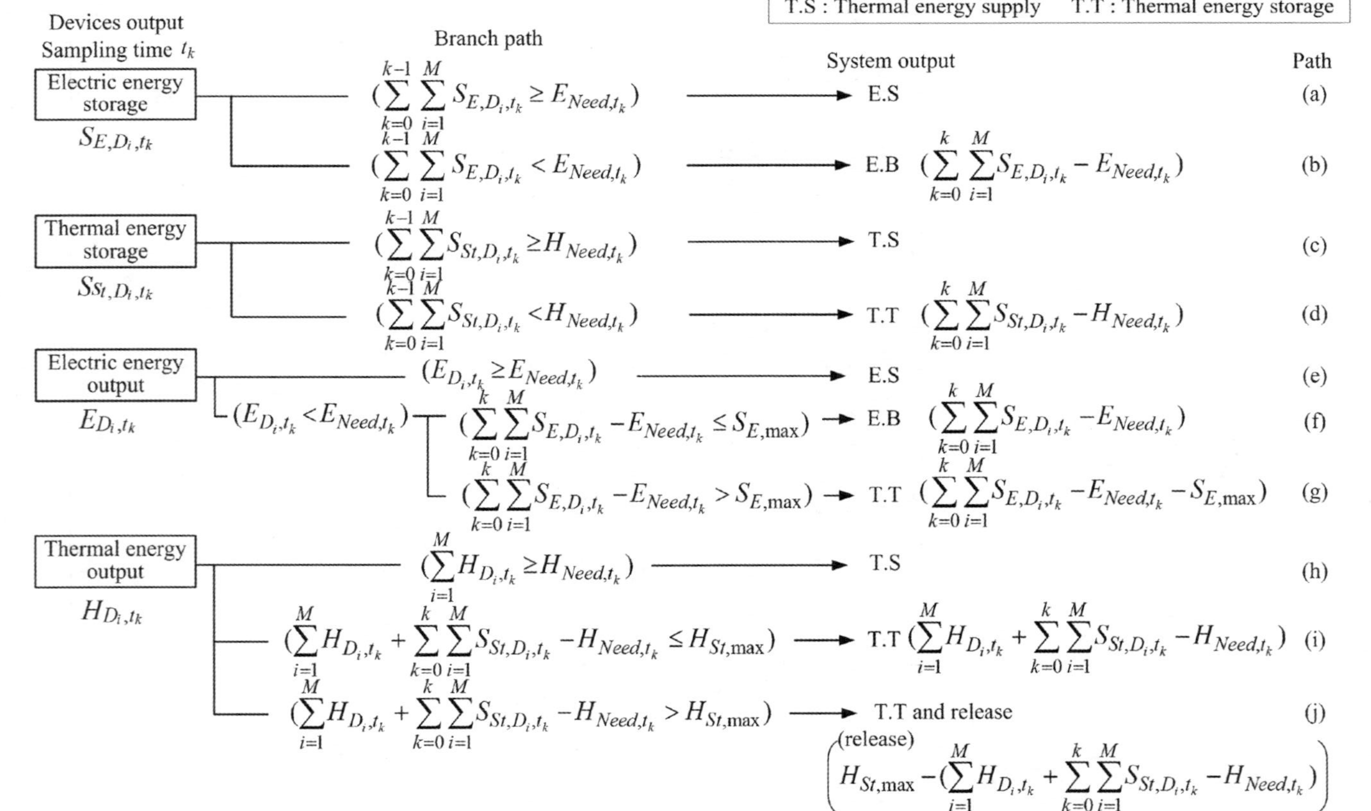

Figure 4.7 Energy supply path.

consumption) of the energy equipment, (3) Minimization of the carbon dioxide gas emission accompanying operation, and (4) Minimization of the three objective functions described above. Equations (4.7) and (4.8) are energy-balance equations of electric power and heat, respectively,

$$E_{FS,t_k} + E_{Utility,t_k} + E_{Stp,t_k} = E_{System,t_k} + \Delta E_{EL,t_k} + \Delta E_{HP,t_k} + \Delta E_{CPH,t_k} + \Delta E_{CPO,t_k} + E_{H,t_k} \quad (4.7)$$

$$\alpha_{FS} \cdot F_{FS,t_k} \cdot \phi_{FS} + \alpha_{Boiler} \cdot F_{Boiler,t_k} \cdot \phi_{Boiler} + H_{HP,t_k} + H_{St,t_k} = H_{System,t_k} + H_{Rad,t_k} + \Delta H_{St,t_k} \quad (4.8)$$

The left-hand side in eqns (4.7) and (4.8) is the amount of energy input into the system, and the right-hand side expresses the amount of energy output from the system. Here, E_{FS,t_k}, $E_{Utility,t_k}$, E_{Stp,t_k} and E_{System,t_k} express the electric power of fuel cell stack, commercial power and power storage, respectively. $\Delta E_{EL,t_k}$, $\Delta E_{HP,t_k}$, $\Delta E_{CPH,t_k}$, $\Delta E_{CPO,t_k}$ and E_{H,t_k} express the power consumption of the electrolyzer, heat pump, hydrogen compressor, oxygen compressor and heater, respectively. α_{FS} and α_{Boiler} express calorific value of fuel of fuel cell stack and boiler. ϕ_{FS} and ϕ_{Boiler} express efficiency of fuel cell stack and boiler. F_{FS,t_k} and F_{Bolier,t_k} express the fuel flow of the fuel cell stack and boiler, respectively. H_{HP,t_k}, H_{St,t_k}, H_{System,t_k}, H_{Rad,t_k} and $\Delta H_{St,t_k}$ express the heat of heat pump, heat-storage tank, system, radiator and heat storage loss, respectively. Objective function (1) described in Section 4.1 is an operating pattern when the difference in input-output of the energy balance eqns (4.7) and (4.8) serves as the minimum. Objective function (2) is an operating pattern when the fuel cost and commercial power cost serve as the minimum. The operation cost of equipment D_i between time t_k and Δt is calculated from the fuel flow rate F_{D_i,t_k} and unit fuel price C_{fuel,D_i} that are supplied to the equipment. Therefore, the operation cost of the whole system is calculated by eqn (4.9). Here, $C_{Utility,t_k}$ express the commercial power cost,

$$C_{System,t_k} = \sum_{i=1}^{M} \left(C_{fuel,D_i} \cdot F_{D_i,t_k} \cdot \Delta t \right) + C_{Utility,t_k} \quad (4.9)$$

Objective function (3) expresses the operation pattern whose amount of greenhouse-gas discharge calculated from the fuel consumption is the minimum. Amount of emission W_{System,t_k} of greenhouse gases is calculated by eqn (4.10). However, the number of gas compositions that contribute to greenhouse-gas discharged by equipment D_i is expressed by S.

$$W_{System,t_k} = \sum_{i=1}^{M} \sum_{j=1}^{S} \left(G_{D_i,EX_j} \cdot \varepsilon_{D_i,EX_j,t_k} \cdot F_{D_i,t_k} \cdot \Delta t \right) \quad (4.10)$$

Here, G_{D_i,EX_j} expresses a global-warming factor per unit weight of fuel, $\varepsilon_{D_i,EX_j,t_k}$ being the weight concentration of EX_j, and F_{D_i,t_k} being the amount of

Table 4.2 Energy cost and greenhouse-warming coefficient (Japanese Environment Agency, 2000).

Kerosene fuel	0.01097 Dollar/J 3.099 kg CO_2/kg 2.026 kg/Dollar
Methanol fuel	0.01772 Dollar/J 1.379 kg · CO_2/kg
Commercial power	0.0647 Dollar/J (9:00–21:00) 0.01515 Dollar/J (22:00–8:00) 0.000099167 kg CO_2/kJ

fuel supply to equipment D_i. Table 4.2 shows fuel cost and a global-warming factor,[10] and is analyzed using these values in an analysis case.

4.3.2 Multiobjective Optimization

As shown in eqn (4.11), the operation pattern that minimizes the sum that multiplies each objective function by weight is a multiple-objective optimal solution,

$$\text{minimize}\left(\sum_{t_k=1}^{Period}\sum_{j=1}^{N_p} w_j \cdot f_j(x_{t_k})\right) \tag{4.11}$$

Because it is necessary to compare the evaluation item (energy loss, operation cost, the amount of greenhouse gases discharged) of the system, each evaluation item is converted into the amount of kerosene used in this chapter. w_j in eqn (4.11) is given beforehand and the value of this equation searches for the minimum solution using GA. Here, w_j, $f_j(x_{t_k})$ and N_p express the weight of an objective function, objective function and number of objective functions, respectively.

4.4 Analysis Results

4.4.1 Results of Optimization

Tables 4.3 and 4.4 are the calculation results when optimizing under each objective function using the energy-demand pattern of a representative day in February and August, and convert all values into a kerosene weight. The number in brackets in the table is the value of the conventional energy system (using commercial power and a kerosene boiler). If this system is optimized by operation cost minimization, compared with the cost of the conventional system, there will be a maximum of 4% and 16% reduction in February and August, respectively. Reduction rates differ every month because the energy demanded and the solar module output are different.

Table 4.3 Calculation result of each objective function in February.[a]

Minimization of	*Operation cost*	*The error of demand-and-supply balance*	*Greenhouse gas*
Minimization of operation cost	**14.72** (15.36)	0.439	13.76 (13.35)
Minimization of the error of demand-and-supply balance	22.40	**0.0170**	18.82
Minimization of the amount of greenhouse gas discharge	15.66	0.426	**13.16**

[a]values shown as kerosene (kg).

Table 4.4 Calculation result of each objective function in August.[a]

	Operation cost	*The error of demand-and-supply balance*	*Greenhouse gas discharge*
Minimization of operation cost	**3.61** (4.28)	0.795	3.42 (2.60)
Minimization of the error of demand-and-supply balance	6.85	**0.0247**	4.174
Minimization of the amount of green-house gas discharge	5.27	0.199	**2.55**

[a]values shown as kerosene (kg).

Table 4.5 RMC in February.[a]

	Commercial power	*Fuel cell*	*Solar module*		*Heat pump*	*Boiler*
			Electricity use	Heat use		
Minimization of the error of demand-and-supply balance	0.90	0.906	0.338	0.662	0.980	0.964
Operation cost minimization	0.57	0	0.361	0.639	0.982	0.990
Greenhouse gas minimization	0.82	0.899	0.207	0.793	1.00	1.00

[a]RMC = Maximum output/Device capacity.

4.4.2 Equipment Capacity

The analysis results of the ratio of maximum output to equipment capacity at the time of planning operation with each objective function (this value is

Table 4.6 RMC in August.

	Commercial power	*Fuel cell*	*Solar module*		*Heat pump*	*Boiler*
			Electricity use	Heat use		
Minimization of the error of demand-and-supply balance	0.87	0.714	0.159	0.841	0.604	0.255
Operation cost minimization	0.81	0.631	0.566	0.434	0.614	0.564
Greenhouse gas minimization	0.48	0.503	0.444	0.556	0.890	0.139

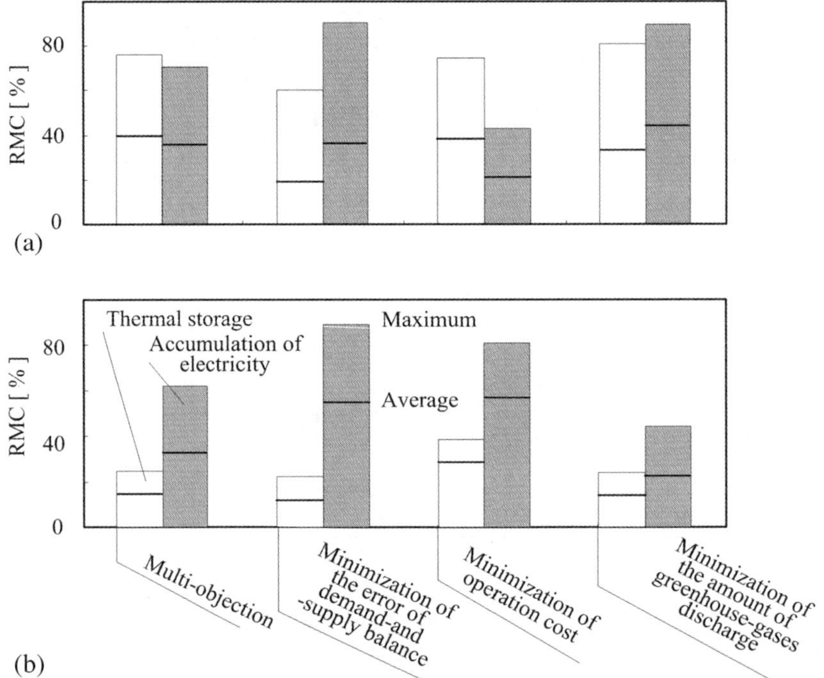

Figure 4.8 Energy storage result; (a) February; (b) August.

described as RMC below) are shown in Tables 4.5 and 4.6. If the value of RMC is lower than 1, a decrease in the initialized equipment capacity (Table 4.1) is possible. On the other hand, equipment with the value of RMC greater than 1 has insufficient capacity. The analysis results of RMC of heat storage and power storage (storage of hydrogen and oxygen by water electrolysis) when operating the system under each objective function are shown in Figure 4.8.

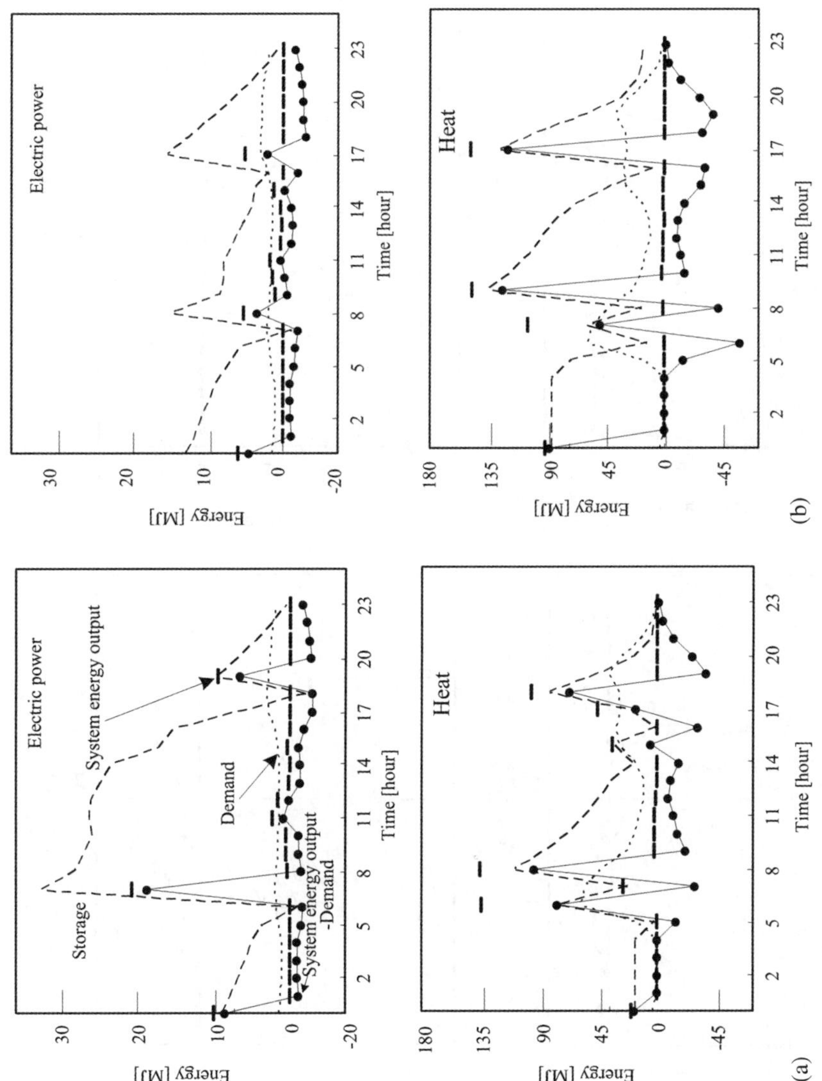

Figure 4.9 Operational planning result of February; (a) Minimization of the error of demand-and-supply balance; (b) Minimization of operation cost; (c) Minimization of the amount of greenhouse-gas discharge; (d) Multiple-purpose operational planning result.

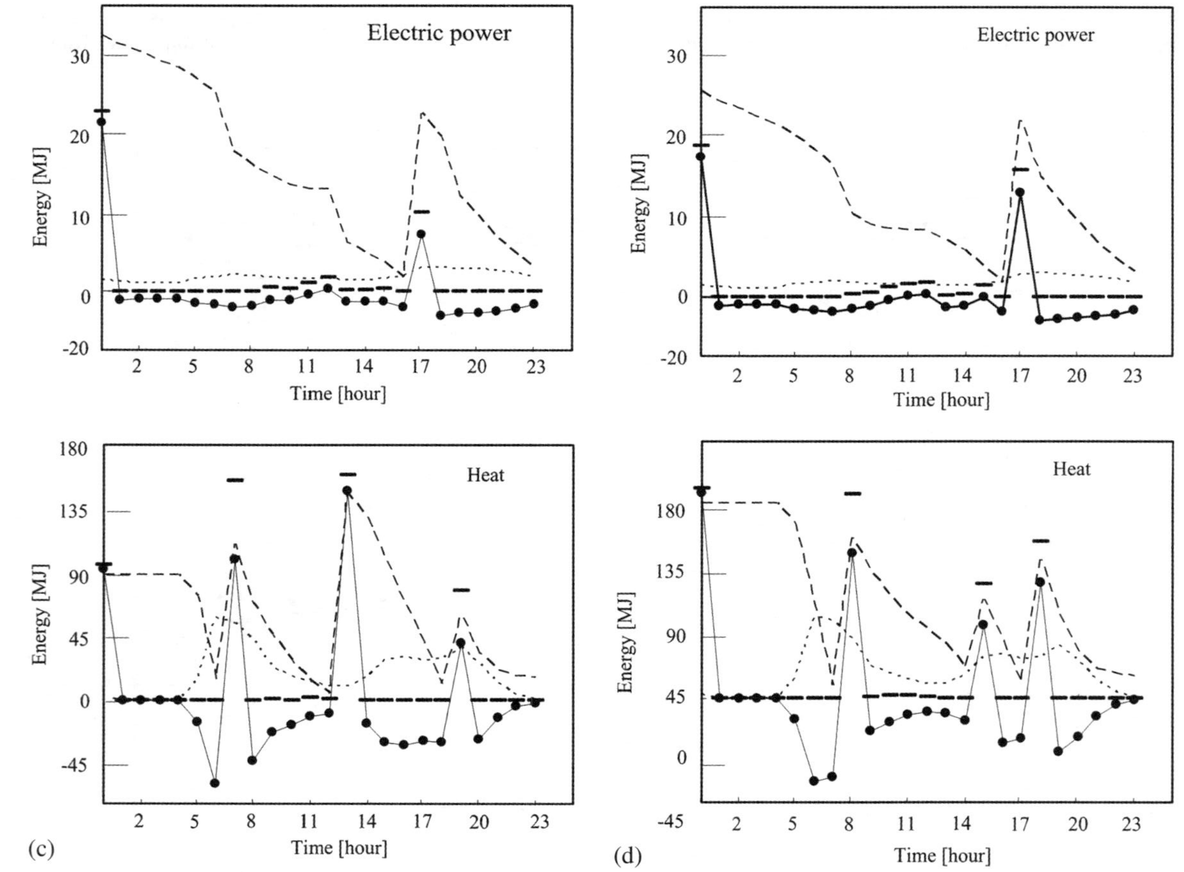

Figure 4.9 Continued.

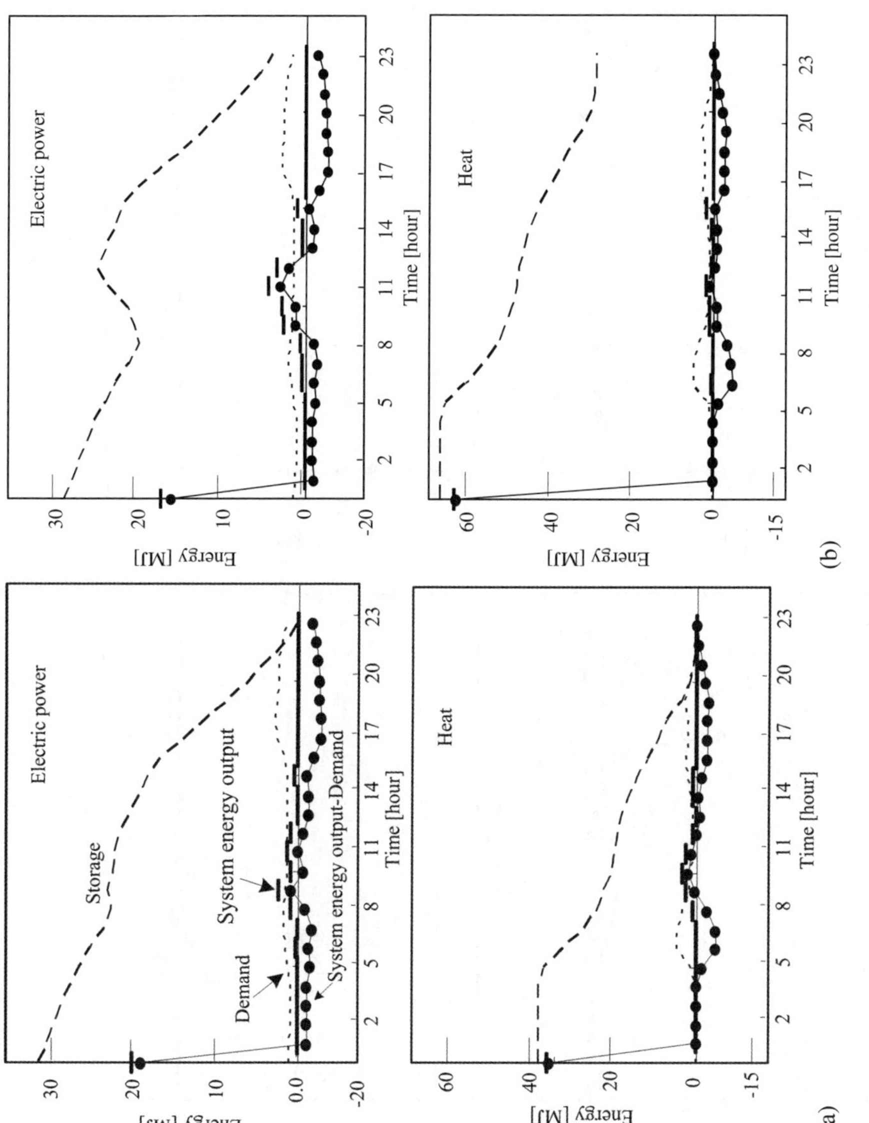

Figure 4.10 Operational planning result of August; (a) Minimization of the error of demand-and-supply balance; (b) Minimization of operation cost; (c) Minimization of the amount of greenhouse-gas discharge; (d) Multiple-purpose operational planning result.

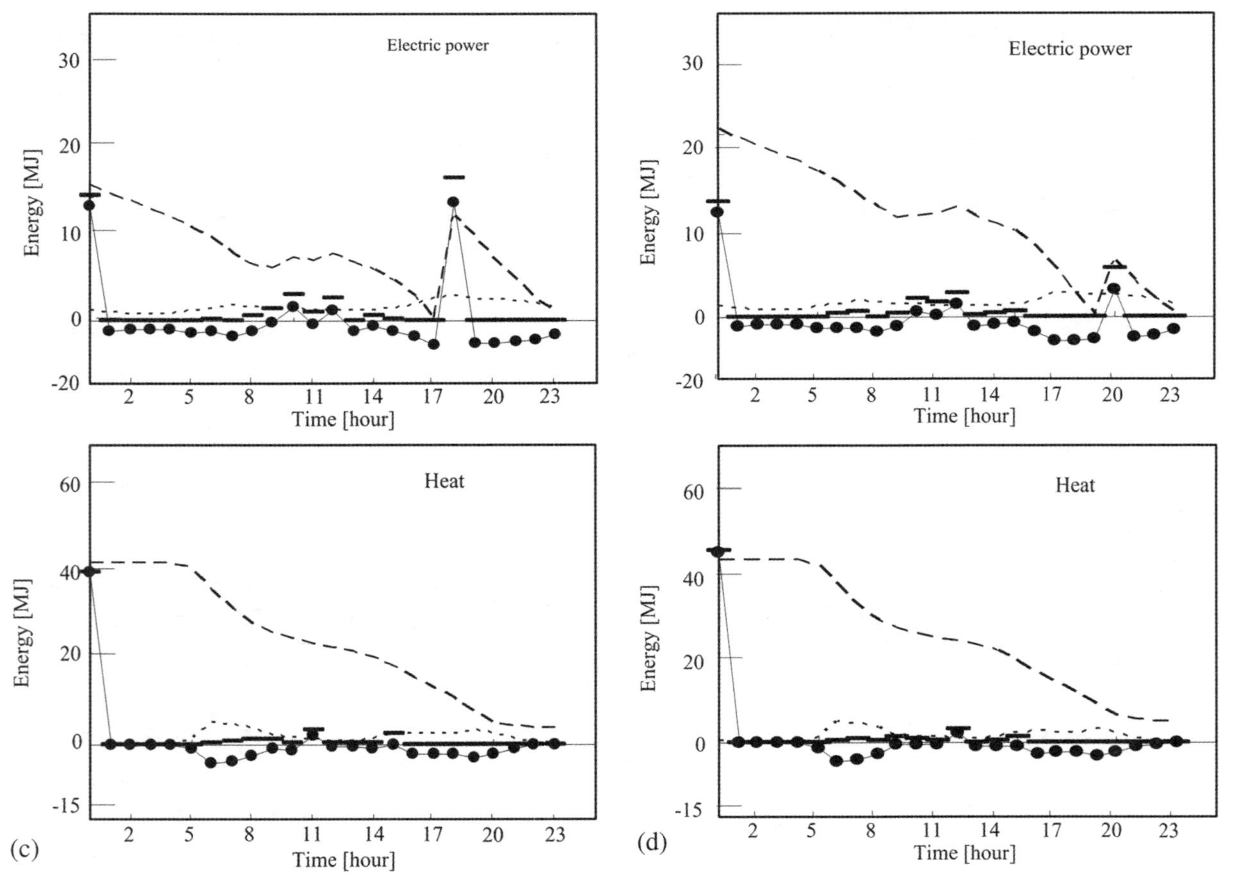

Figure 4.10 Continued.

From the results of Figure 4.8, the capacity-reduction ratio of a heat-storage tank and power storage equipment can be seen. From the method described above, the optimization of the capacity of each piece of equipment that composes a system can be designed.

4.4.3 Objective Function and Characteristics of Operation Plan

Figures 4.2 and 4.3 show the balance results of electric power and heat when optimizing an operation plan under each objective function using the energy-demand pattern of representative February and August days. "System energy output" in these figures is the characteristic of electric power and heat outputted by the system configuration equipment. However, power-storage output and heat-storage output are not included in this characteristic, but are separately shown as "Storage". Moreover, the energy demanded is shown as "Demand".

In Figures 4.9 and 4.10, although the characteristic of "Demand" is flat, there are times when the characteristic of "System energy output" is extremely large. The reason for this is that it stores energy for a short period in the system, when the stored energy is released in large quantities. Figure 4.11 shows the ratio of energy output by each piece of equipment. The operation plan when optimizing the energy system of Figure 4.1 under different objective functions from the analysis results described above has the characteristics described below.

(1) Operation plan of the minimization of error in demand-and-supply balance.

As Figures 4.9(a) and 4.10(a) show, the stored electric power and heat is mostly used up at 23:00. Moreover, as shown in Tables 4.5 and 4.6, both months are high-output months, and the operating point of a fuel cell is planned so that partial-load operation is avoided. The operation of a fuel cell with a high value of RMC avoids low-efficiency, partial-load operation.

(2) Operation plan of the minimization of equipment operation cost (fuel consumption).

From the results of Tables 4.5 and 4.6, the electric power generated by the solar module is supplied to the demand side with electric power rather than heat compared with other objective functions. This is because the cost of this method of production is high compared with the amount of heat. As the results of Table 4.5 show, the output of commercial power reduces and there is no fuel cell operation. For optimization of this objective function, priority is given to the operation plan in order of; suspend operation, convert power into heat, and store power.

(3) Operation plan of the minimization of the greenhouse-gas emission.

From the results of Table 4.5, operation of a heat pump and boiler is planned with a maximum-output point on representative February days with great heat demand. Moreover, application is planned with a value with a high

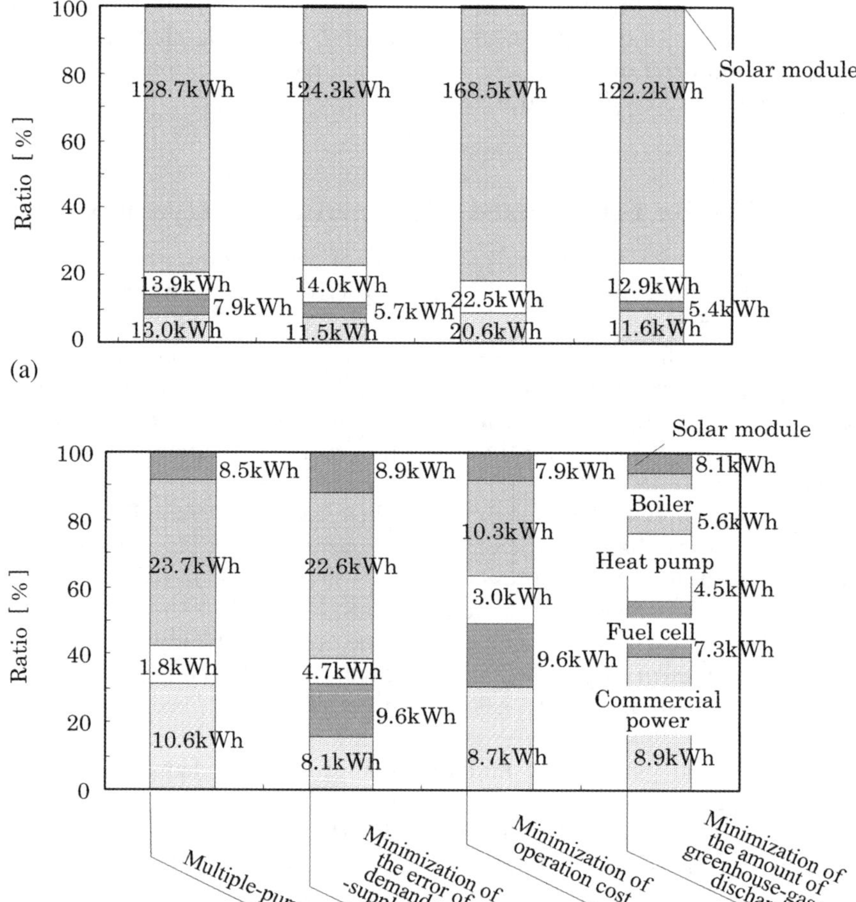

Figure 4.11 Result of energy output component; (a) February representative day; (b) August representative day.

supply rate by the heat of the solar module, and high RMC of the fuel cell. On the other hand, heat supply, rather than the boiler, is mainly concerned with the heat pump according to the results of representative August days with little heat demanded, as shown in Table 4.6 and Figure 4.11(b).

(4) Operation plan of the minimization of multiobjective functions.

When operating the system under a multiobjective function, the result of Figure 4.8 to RMC is a value near the average value adding each objective function. As shown in Figures 4.9(d) and 4.10(d), the operation plan under this objective function includes the characteristic of the operation plan of other objective functions.

4.5 Conclusions

How the operation optimization of the combined system of PEFC-CGS, a solar module, a geothermal heat pump, heat storage and power storage equipment, a commercial electric power, and a boiler using GA should be analyzed was described. The capacity optimization of equipment that composes a system was considered from the analysis results. If the capacity of each energy device changes with the objective functions given to the system and examination of equipment cost is added to the results in this chapter, it can be utilized in real design. Moreover, the characteristics when planning the operation of a system under each objective function were investigated. As a result, for example, the minimization objective of operation cost, operation that suspends operation and converts power into heat, power storage operation is planned. If the objective function concerning minimization of greenhouse-gas discharge is given to the system, there will be many opportunities to use the electric power of a solar module as heat; moreover, heat supply mainly concerned with the heat pump rather than the boiler is planned. In the design and operation plan of a combined system containing renewable energy equipment, care should be taken concerning the composition and operating method that change greatly according to objectives given to the system.

Acknowledgments

This work was partially supported by a Grant-in-Aid for Scientific Research(C) from the JSPS.KAKENHI(17510078).

Nomenclature

C: cost in dollars
D_i: energy equipment (i the number of pieces of equipment)
E: electric power kW
E_c: chemical equivalent
E_V: voltage V
ΔE: electric energy consumption kW
F: fuel quantity of flow kg/s
F_d: Faraday constant (96 500 C/mol)
F_m: objective function
G_{D_i,EX_j}: greenhouse warming coefficient per unit weight of the gas EX_j discharged from equipment D_i kg/kg
H: heat kW
ΔH: heat consumption kW
I: current A
M: number of the composition equipment of a system
N_{dv}: number of individuals of a chromosome model
N_p: number of objective functions
p: probability

Q_{f,t_k}: hydrogen amount of supply mol/s
Q_{H2,t_k}: amount of hydrogen produced between Δt from the sampling time t_k kg
R: operating period of a system s
S: number of composition of the exhaust gas that contributes to greenhouse warming
S_E: power quantity to be stored kJ
S_{St}: amount of heat stored kJ
t_k: sampling time s
Δt: interval time s
W: amount of greenhouse-gas discharge kg/s
ΔW_{FS}: power loss of a cell stack kW
w_j: weight of an objective function
$F_j(x_{t_k})$: objective function

Greek Symbols

α: calorific value of fuel kJ/kg
ε_{D_i,EX_j}: weight concentration of the gas EX_j discharged by Equipment D_i kg/kg
ϕ: efficiency %
ρ: density of heat-storage medium kg/m^3

Subscripts

CPH: hydrogen compressor
CPO: oxygen compressor
EL: electrolyzer
EX: emission gas
FS: fuel cell stack
H: heater
HP: heat pump
Rad: radiator
St: heat-storage tank
Stp: power storage
Utility: commercial power

References

1. K. Ito, R. Yokoyama, *The optimal program of cogeneration*, 1990, Sangyotosyo Inc, Tokyo, 25–43 (in Japanese).
2. K. Ito, T. Shibata and R. Yokoyama, Optimal operation of a cogeneration plant in combination with electric heat pumps, *Trans. ASME J. Energy Res. Technol.*, 2002, **116**, 56–64.

3. D. E. Goldberg, *Genetic Algorithms in Search Optimum Machine Learning,* 1989, Addison Wesley Toronto, Canada.
4. K. Narita, *The research on unused energy of the cold region city and utilization for the district heat and cooling,* 1996, Ph.D. thesis; Hokkaido University (in Japanese).
5. S. Obara, K. Kudo and K. A. Ismail, Study on small-scale fuel-cell co-generation system with methanol steam reforming consideration partial load and load fluctuation, *Trans. JSME Series B*, 2003, **677**(69), 139–147 (in Japanese).
6. S. Obara and K. Kudo, Operational Optimization and Scheduling of Multikind Small Capacity Energy Devices for Cold-Region Houses, *Proceedings of 9th International Thermal Energy Storage*, 2003, **11**, 297–302.
7. HC-TECH Inc. HC12a and HC22a properties and performance tests data sheets, 1997, 1–3.
8. A K. Colling, R J. Roy, Development status and testing of high differential pressure SPE water electrolysis cells, SAE Technical Paper, 1998, SAE-981802, 1–8.
9. K. Nagano, T. Mochida, K. Shimakura, K. Murashita and S. Takeda, Development of thermal-photovoltaic hybrid exterior wallboards incorporating PV cells in and their winter performances, *Sol. Energy Mater. Sol. Cells*, 2002, **77**, 265–282.
10. The investigative commission about the amount calculation method of greenhouse-gas discharge, *Enforcement ordinance discharge coefficient list*, 2000, Japanese Environment Agency (in Japanese).

CHAPTER 5

PEFC/Engine Generator Compound Energy System (1) – CO_2 Discharge Characteristic of PEFC/Hydrogen-Gas-Engine Hybrid Cogeneration

SHIN'YA OBARA

5.1 Introduction

Generally, a proton-exchange membrane fuel cell (PEFC) has the maximum power-generation efficiency, higher than an engine generator. In using a fuel cell in city areas, such as a normal house, low operations with load factor (load added to a fuel cell/fuel cell capacity) occur frequently due to load fluctuation. A fall in load factor will reduce the power-generation efficiency of a fuel cell. Therefore, in order to introduce power plants, such as fuel cells, into a city area, it is necessary to consider power-generation-efficiency characteristics at a time of partial load. The power-generation efficiency of PEFC needs to synthesize and evaluate auxiliary machinery concerning reformed gas, such as the reformer, cell stack, and carbon-monoxide oxidization equipment, as well as the auxiliary machinery of electric systems, such as the inverter. The power-generation efficiency of PEFC at a time of partial load influences the reformer structure, cell stack structure, the heat-conducting characteristics and the

RSC Energy Series No. 3
Compound Energy Systems: Optimal Operation Methods
By Shin'ya Obara and Arif Hepbasli

Published by the Royal Society of Chemistry, www.rsc.org

temperature-control characteristics of each reaction part. The structure of the reformer and cell stack affects the reaction efficiency of each part. Moreover, the reaction of each part is dependent on the temperature conditions of the system. For this reason, reaction efficiency also changes with the heat-conducting characteristics and the temperature-control methods of each part. Therefore, it is predicted that power-generation-efficiency characteristics at a time of partial load will change with the construction of each part of PEFC under consideration. In this way, the fall in power-generation efficiency and an increase in carbon-dioxide emissions at a time of partial load are a subject of a PEFC. The objective of this study is to improve these. The investigation concerning an improvement of the power-generation efficiency at a time of the low load of a fuel cell system has not been found in previous studies. On the other hand, hydrogenation technology has been studied concerning the improvement in efficiency at the time of exhaust gas cleanup and the partial load of a gas engine.[1,2] From these research findings, improvement of exhaust cleanup and brake thermal efficiency is checked by increasing the hydrogen rate of fuel at a time of low load.[3–6] Below, a hydrogen-gas engine is described to be NEG. So, in this chapter, the fuel consumption and the carbon-dioxide emission characteristics of a PEFC/NEG hybrid cogeneration system (HCGS) are investigated concerning power-generation efficiency improvement at a time of partial load. Since the history of engine technology is long, NEG cogeneration is reliable, and it is cheap compared with the fuel cell. However, compared with PEFC, it is inferior, especially in respect of the maximum power-generation efficiency and noise. So, in this chapter, the method of operating PEFC corresponding to base load, and following NEG at load fluctuation and operating is proposed. Compared with when NEG or PEFC is operated independently (IEG, IPF), the facility costs of HCGS are high. Therefore, in order for the introduction of HCGS to city areas to be effective, compared with IEG or IPF, power costs and carbon-dioxide emissions need to be advantageous. In this chapter, the capacity of fuel consumption, carbon-dioxide emissions, the boiler, and the heat-storage tank are investigated by numerical analysis regarding the introduction of HCGS into apartment houses in Sapporo in north Japan, which is a cold, snowy area.

5.2 System Scheme

5.2.1 HCGS Model

A block diagram of HCGS proposed in this chapter is shown in Figure 5.1(a). There is a system by which city gas is supplied to NEG, to the reformer, to the heat-source burner, and to the boiler. Therefore, the proposed amount of city gas consumption of the system is calculable by

$$U_{total,t} = U_{N,t} + U_{R,t} + U_{B,t} + U_{S,t} \tag{5.1}$$

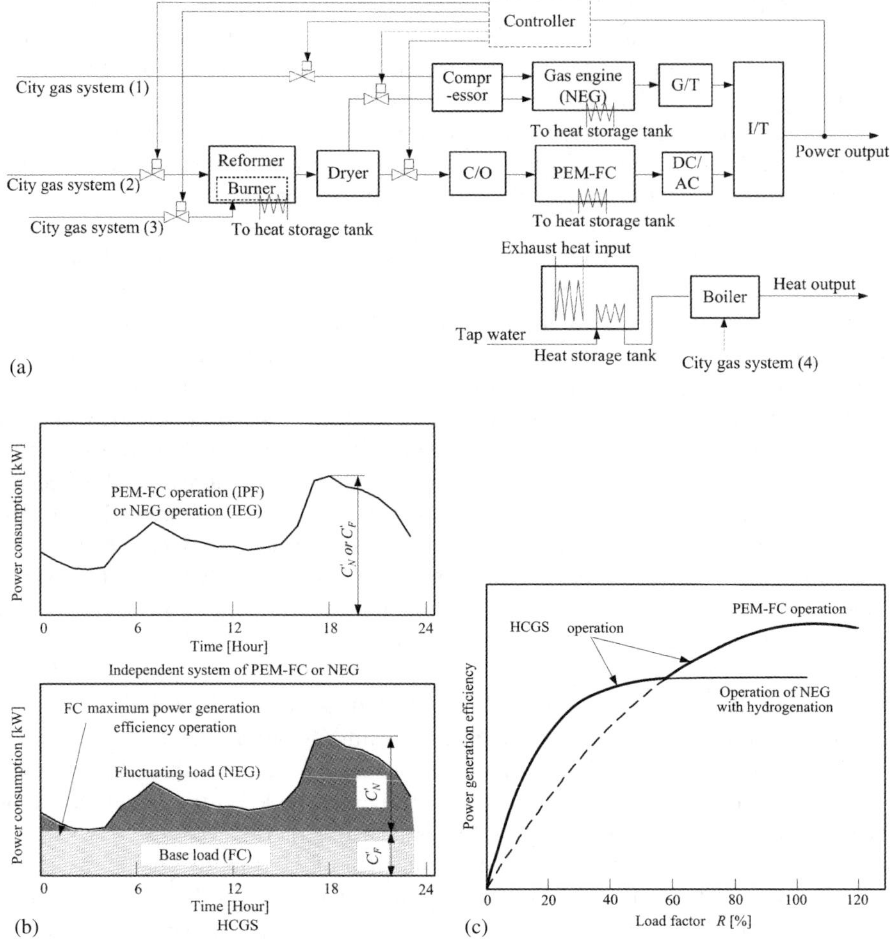

Figure 5.1 HCGS configuration; (a) HCGS block diagram; (b) System operation; (c) Efficiency characteristic model of HCGS.

In order to simplify the analysis, city gas is set at CH_4. Equation (5.2) is a reaction that changes city gas into reformed gas with a high hydrogen concentration,

$$CH_4 + H_2O \rightarrow CO + 3H_2 - 206 \quad [kJ/mol] \tag{5.2}$$

Since there is a lot of water in reformed gas, after removing some of the water by a dryer, PEFC and NEG are supplied. There is a capacity ratio of about 80% of hydrogen in the reformed gas after removal of the water. Moreover, 1% or less of carbon monoxide is contained in reformed gas. Carbon monoxide reduces the power of the fuel cell. So, when supplying reformed gas to PEFC, the carbon-monoxide concentration is reduced to several ppm using the reaction of eqn (5.3) through carbon-monoxide oxidization equipment,

$$CO + H_2O \rightarrow CO_2 + H_2 + 41 \quad [kJ/mol] \tag{5.3}$$

Since eqn (5.2) is an endothermic reaction, it burns city gas by the burner installed in the reformer, and is taken as a heat source. Equation (5.4) is the burning reaction of city gas,

$$CH_4 + 2O_2 \rightarrow CO_2 + 2H_2O + 802 \quad [kJ/mol] \tag{5.4}$$

The reactor size of a reforming reaction is assumed based on results in the literature.[7–9] The power of PEFC is changed into the regulation frequency (50 Hz) of an alternating current of 100 V with a DC–AC converter and an inverter. City gas and reformed gas can be mixed and supplied to NEG. The power generator linked to NEG is a single-phase synchronous type, and controls the frequency with an inverter like a fuel cell. The quantity of reformed gas supplied to PEFC, and the quantity of city gas and reformed gas that are supplied to NEG are controlled by the command of the controller. The value of the power load is input into the controller, and the flow rate of the city gas and reformed gas is controlled from the volume of the load.

5.2.2 Compression of Reformed Gas

When supplying reformed gas to the combustion chamber of NEG, it is necessary to raise the pressure of the reformed gas with a compressor. Consequently, compression work $L_{c,rg,t}$ of the reformed gas is assumed to be compression work by an ideal gas, and is calculated by

$$L_{c,rg,t} = P_{\infty} \cdot U_{\infty,t} \cdot \ln(U_{rg}/U_{\infty})/\eta_c \tag{5.5}$$

t is sampling time and η_c is the overall efficiency of a compressor. A cooling loss, mechanical loss, and other losses are included in the overall efficiency of a compressor. The compressor overall efficiency of a real system is about 0.6. Moreover, reformed gas is pressurized with a compressor at 0.1 MPa. The reformed gas heating value supplied to the engine combustion chamber is set at $q_{rg,t}$, the efficiency of the power generator is set up with η_G, and the power output by NEG is calculated from

$$E_{N,t} = \eta_G \cdot q_{rg,t} - L_{c,rg,t} \tag{5.6}$$

5.2.3 Operating Method of System

Figure 5.1(b) shows the operation model by PEFC or NEG, and the operation model of HCGS. The system that introduces PEFC independently is described as IPF, and the system that introduces NEG independently is described as IEG. C'_N

and C'_F of this figure express the maximum load of NEG and PEFC, respectively. It is necessary to arrange the capacity (C_N and C_F) of NEG and PEFC introduced into each operation model with a value exceeding C'_N and C'_F. In the operation model of HCGS, since PEFC is used for a base load, FC can be operated in a high-load region where efficiency is high. For load regions other than base load where low load frequently occurs, it corresponds to NEG. The following effects are expected from the operation method of HCGS.

(a) In the low-load operating range of NEG, an improvement in exhaust cleanup and brake thermal efficiency is expected by increasing the hydrogen concentration of the fuel. Therefore, improvement in power-generation efficiency is expected by corresponding to NEG in a load fluctuation region where low load frequently occurs.
(b) Since PEFC is operated at constantly high load, the power-generation efficiency of the whole system may improve.
(c) Since PEFC and NEG are introduced into HCGS, the facility is complicated compared to IEG or IPF. However, the capacity NEG and PEFC, which are introduced into HCGS, are small compared with IEG or IPF. Therefore, HCGS may be advantageous even if there is a trade-off relationship with the rise in facility costs for improvement in power-generation efficiency.

5.2.4 Power-Generation-Efficiency Characteristics of HCGS

Figure 5.1(c) is the model showing the relation between NEG, the load factor of PEFC, and power-generation efficiency. In the low-load operating range of NEG, an improvement in brake thermal efficiency is expected by increasing the hydrogen concentration of the fuel.[3] However, there is no effect of hydrogenation in the high-load operating range of NEG. Moreover, as shown in Figure 5.1(c), compared with PEFC, the maximum efficiency of NEG is low. Therefore, it is expected that operation of NEG in a low-load region is advantageous, and operation of PEFC in a high-load region is advantageous. From the characteristics of the power-generation efficiency of PEFC and NEG described above, the operation method of HCGS of Figure 5.1(b) is the most advantageous in power-generation efficiency in the whole system compared with IEG or IPF.

5.3 Equipment Characteristics

5.3.1 Output Characteristics of NEG

(1) Output characteristics.

The output characteristics of a city gas engine with hydrogenation introduce the evaluation results studied in the past. The power generator assumed to have the engine specifications examined is shown in Table 5.1. Figure 5.2(a) shows the hydrogenation rate of the gas engine of Table 5.1(a), and the examination results of brake thermal efficiency.[3] Regions where the engine has an effective pressure of less than 0.8 MPa have effective hydrogenation. On the other hand, even if the mean effective pressure does not add hydrogen in regions exceeding 0.8 MPa, high thermal efficiency can be obtained. Figure 5.2(b) shows the relation between the hydrogenation rate, mean effective pressure, and brake thermal efficiency.[3] The broken line shown in this figure is the hydrogenation rate showing maximum thermal efficiency. Figure 5.2(c) shows the relation between the amount of city gas (CH_4) consumption at the time of maximum thermal efficiency, the amount of hydrogenation, and the production of electricity of NEG. The characteristics of Figure 5.2(c) were calculated from the hydrogenation rate based on the examination results of Figures 5.2(a) and (b). In Figure 5.2(c), when the production of electricity exceeds 14 kW, the amount of hydrogenation is zero. This is because high thermal efficiency can be obtained even if there is no hydrogenation over a large range of engine power, as Figure 5.2(a) describes.

(2) Efficiency.

Figure 5.2(d) shows the relation between the load factor (load/capacity) of NEG, and power-generation efficiency. Although reformed gas is supplied to NEG, reformer efficiency is included in the power-generation efficiency shown in Figure 5.2(d). Here, eqn (5.7) defines reformer efficiency,

$$\eta_R = (q_{H_2}/q_{CH_4}) \times 100 \quad [\%] \tag{5.7}$$

q_{H_2} in eqn (5.4) is the calorific power of hydrogen in reformed gas, and q_{CH_4} expresses the calorific power of the city gas supplied to a reformer. In this chapter, η_R was set up at 73%.[7–9] Moreover, in the following analysis example, 10-kW NEG is introduced into IEG and 5-kW NEG is introduced into HCGS. Generally, engine thermal efficiency increases, so that the capacity becomes large. Consequently, as shown in Figure 5.2(d), the NEG power-generation-efficiency

Table 5.1 Specifications of NEG; (a) Gas engine specifications; (b) Generator specifications.

(a)	
Number of cylinders	Single cylinder
Total stroke volume	1600 cc
Net maximum power	88 kW
Maximum brake thermal efficiency	33%
Number of revolutions	3300 rpm
(b)	
	Single-phase synchronized
Generator type	20 kVA
Rated output	100 V
Rated voltage	90%
Efficiency	50 Hz
Frequency	Automatic voltage regulator

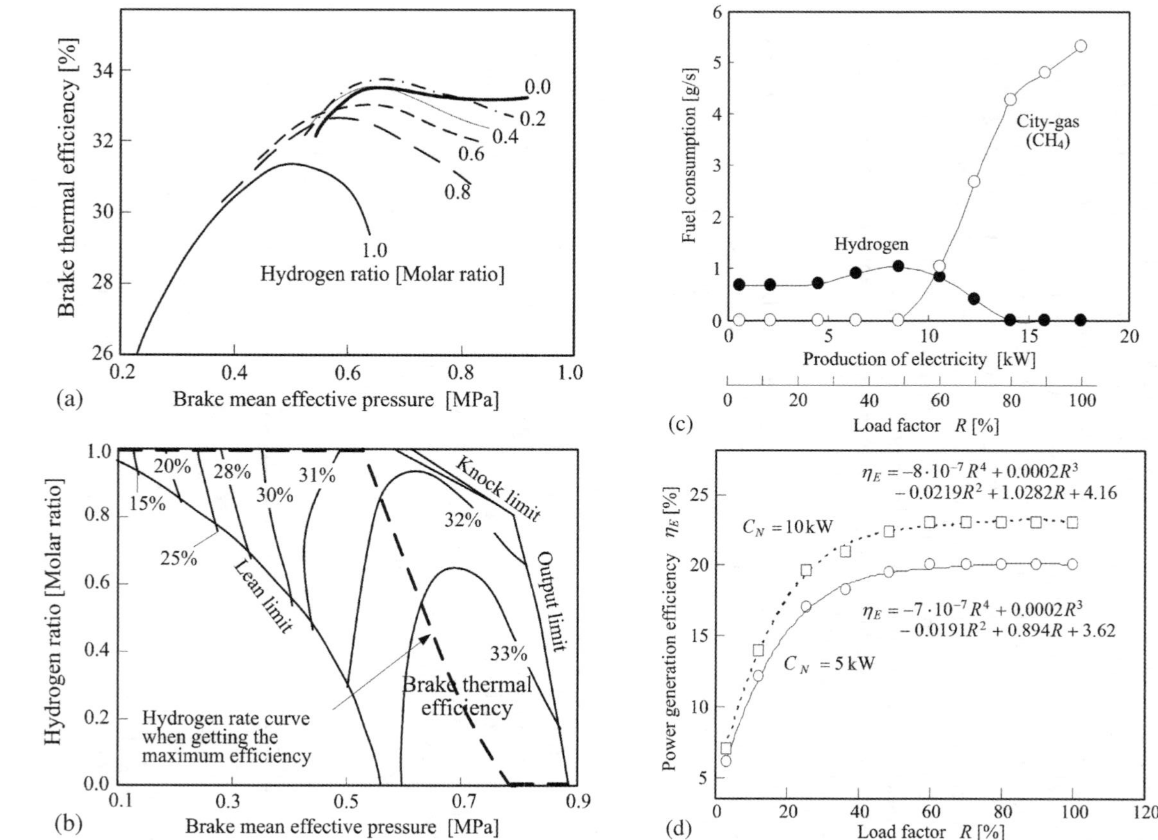

Figure 5.2 Output characteristics of NEG; (a) Hydrogenation rate and thermal efficiency; (b) Amount of optimal hydrogenation; (c) Amount of CH_4 and hydrogenation when getting the maximum thermal efficiency; (d) Relation between production of electricity and gross power-generation efficiency.

characteristics are separated into 5 kW and 10 kW, and analyzed. The difference in the maximum power-generation efficiency shown in Figure 5.2(d) is about 3%. Figure 5.3 shows the results of the power of NEG (5 kW and 10 kW), heat output, and overall efficiency. Heat output includes engine exhaust gas, cooling hot water, and reformer exhaust heat. However, it is assumed that the overall efficiency of Figure 5.3 consumes all power and heat output by NEG. The difference in the overall efficiency of NEG (10 kW and 5 kW) is less than 1%.

(3) Carbon dioxide emissions.

The amount of hydrogen supplied to NEG is determined from the characteristics of Figure 5.2(c). Moreover, CH_4 supplied to a reformer and carbon-dioxide discharged from these reactions is calculable using eqns (5.2)–(5.4). The sampling time is expressed as t and the amount of carbon-dioxide discharged at a reformed reaction at the time of the hydrogenation of NEG is expressed as $g_{N,R,t}$. Moreover, the amount of carbon-dioxide discharged by a heat-source burner is expressed as $g_{N,B,t}$. The amount of carbon-dioxide discharged by the city gas combustion of

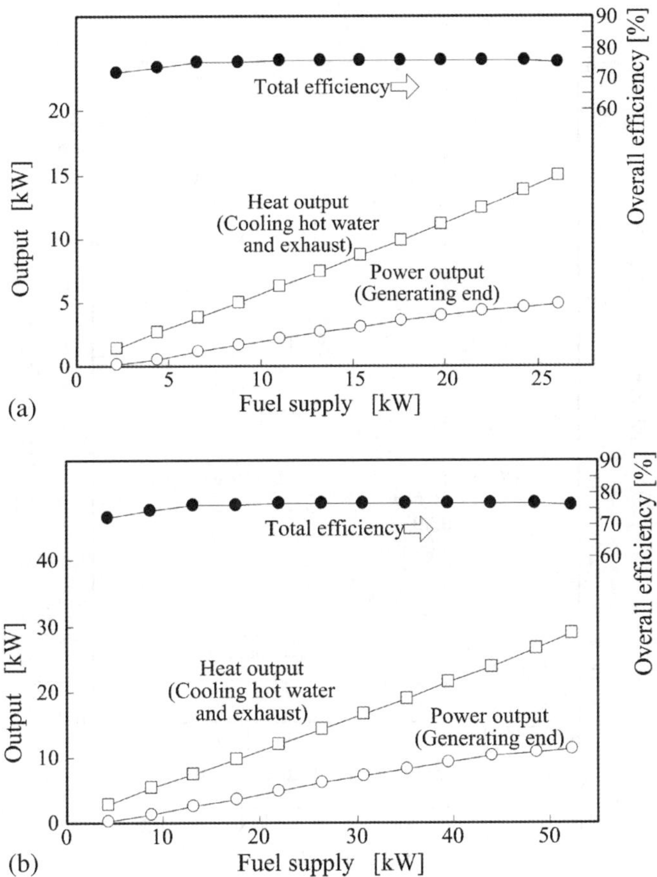

Figure 5.3 Output model of NEG; (a) 5-kW NEG; (b) 10-kW NEG.

NEG is expressed as $g_{N,I,t}$. The amount of carbon dioxide $g_{N,t}$ discharged by the power generation of NEG is calculated by eqn (5.8). The mixed gas of city gas and reformer output gas is supplied to NEG. Therefore, the carbon-dioxide emissions of NEG are the sum of the amount of emission of the reformer (eqns (5.2)–(5.4)), and the amount of emission due to having supplied city gas to NEG. CO_2 emissions in the case of supplying city gas to NEG are calculable using eqn (5.4).

$$g_{N,t} = g_{N,R,t} + g_{N,B,t} + g_{N,I,t} \tag{5.8}$$

Figure 5.4 shows the CO_2 emission characteristics of NEG of Table 5.1. In this figure, an approximate expression showing the relation between the load factor and CO_2 emission is shown. These approximate expressions differ by about 60% of the load factor. The fuel supplied to NEG differs because there are many hydrogen rates in a low-load region, and the rate of city gas is high in a high-load region. In a low-load region, there are many rates of carbon-dioxide discharged by city gas reforming and the burner for the reformers to produce hydrogen. In a high-load region, the rate of carbon dioxide generated

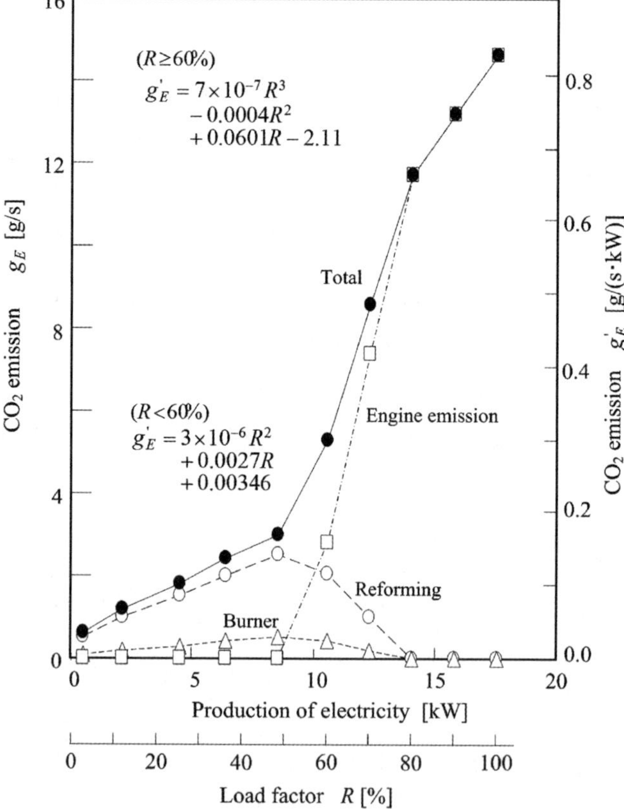

Figure 5.4 CO_2-emission characteristics of NEG.

by burning city gas with an engine is high. The characteristics of Figure 5.4 change with the capacity of NEG, as does the power-generation efficiency. In the calculation of the carbon-dioxide emissions in the case study, 5-kW NEG was set up instead of 10-kW NEG, an increase of 3% at maximum.

5.3.2 Output Characteristics of PEFC

(1) Output characteristics and efficiency.

Figure 5.5(a) shows the output characteristics of 5-kW PEFC with a city-gas reformer as well as a model of overall efficiency.[10,11] The model of Figure 5.5(a) is prepared from examination results and literature on fuel cells.[10,11] Heat output consists of the exhaust heat of the reformer and cell stack. Moreover, the power output is the value of the inverter outlet. Overall efficiency is calculated assuming that all of the power and the heat of a system are consumed. Figure 5.5(b) shows a model of carbon-dioxide emissions and the power-generation efficiency of PEFC of Figure 5.5(a). The carbon-dioxide discharged by the operation of PEFC is through city gas combustion (eqn (5.4)) of the reformer burner and the reforming reaction (eqns (5.2) and (5.3)). Power-generation efficiency η_F of Figure 5.5(b) was calculated using eqn (5.9). It is expressed as $E_{F,t}$ of eqn (5.9); the right-hand side is the power of the inverter outlet of PEFC, the calorific power of CH_4 supplied to a reformer being $q_{R,CH_4,t}$, and the calorific power of CH_4 supplied to the heat-source burner of the reformer being $q_{B,CH_4,t}$. The maximum power-generation efficiency of the fuel cell shown in Figure 5.5(b) is 32%.

$$\eta_{F,t} = \{E_{F,t}/(q_{R,CH_4} + q_{B,CH_4,t})\} \times 100 \tag{5.9}$$

(2) Discharge characteristics of carbon dioxide.

The amount of carbon-dioxide discharged at a reforming reaction is expressed as $g_{F,R,t}$, and the amount of carbon-dioxide discharged by the heat-source burner of a reformer is expressed as $g_{F,B,t}$. The amount of carbon-dioxide discharged $g_{F,t}$ by the power generation operation of PEFC is calculated from eqn (5.10).

$$g_{F,t} = g_{F,R,t} + g_{F,B,t} \tag{5.10}$$

The change rate of the CO_2 emissions of NEG is large in a high-load region, as Figure 5.4 shows. On the other hand, the change in the CO_2 emissions of PEFC is large in a low-load region, as Figure 5.5(b) shows. From the difference in these carbon-dioxide discharge characteristics, the operation of NEG is advantageous in a low-load region, and the operation of PEFC is advantageous in a high-load region.

5.3.3 Carbon-Dioxide Emission Characteristics of Boiler

The amount of carbon-dioxide discharged by the operation of the boiler is calculated using eqn (5.4). Thus, the amount of CH_4, carbon dioxide, and water,

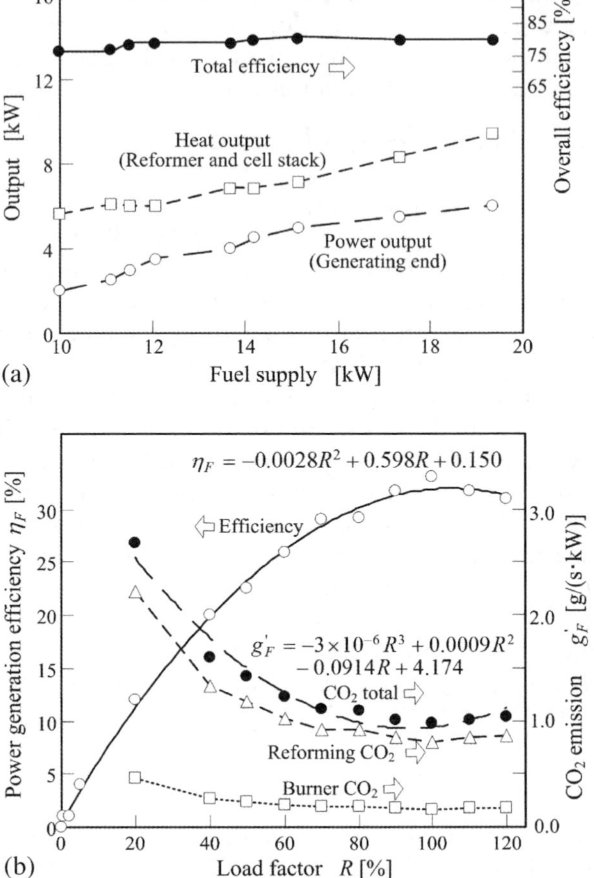

Figure 5.5 Output model of PEFC system; (a) Output characteristics of 5-kW PEFC; (b) The characteristics model of the load factor of a PEFC with reformer, and power-generation efficiency. The area of the electrode including the anode and cathode of the fuel cell stack is 1 m^2, respectively and the reformer efficiency is 73%.

respectively, can be obtained by introducing the heat output required of a boiler into eqn (5.4). However, the boiler efficiency is set at 0.9.

5.4 Power and Heat Output Characteristics of HCGS

5.4.1 System Operation Map

The figure showing the relation between the city gas calorific power supplied to IPF, IEG, and HCGS (amount of consumption) and production of electricity and heat output of a system is defined as an operation map. Figure 5.6(a) shows the operation map of 10-kW PEFC. Moreover, Figure 5.6(b) shows the

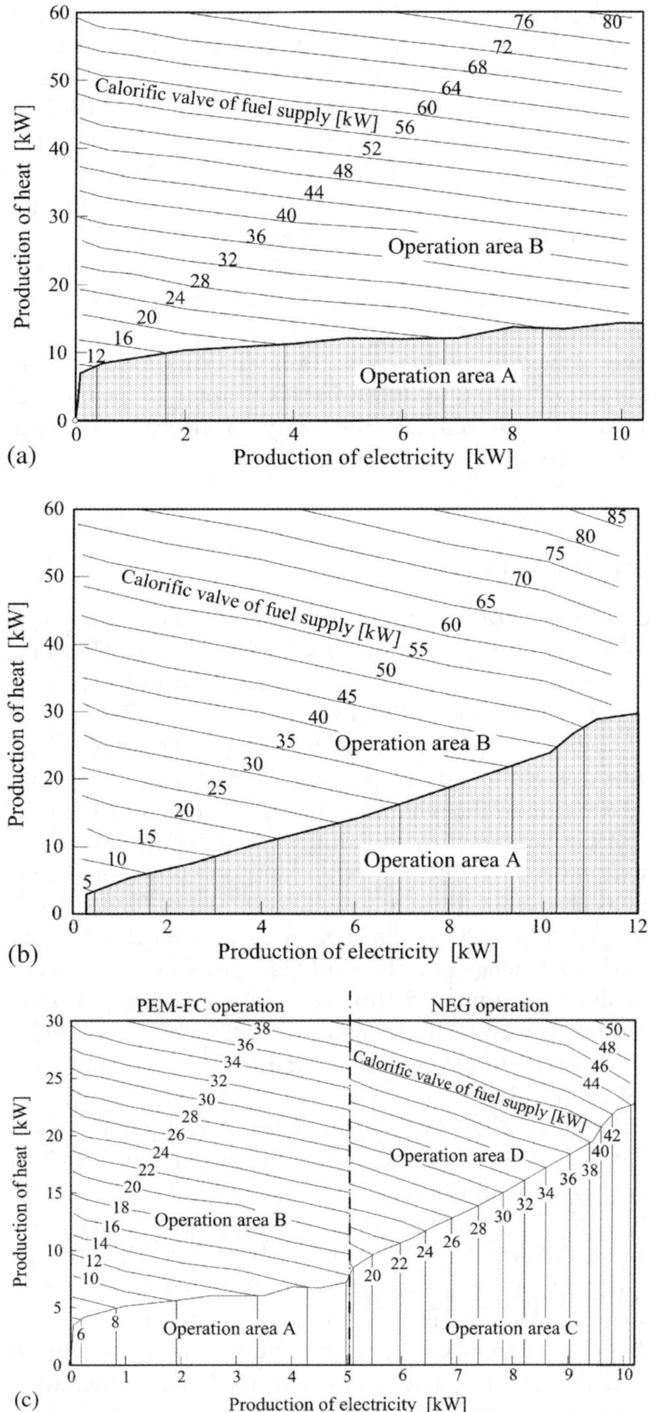

Figure 5.6 HCGS operation map; (a) IPF(10-kW PEFC operation map); (b) IEG (10-kW NEG operation map); (c) HCGS operation map.

operation map of the 10-kW NEG. Operation Area A of each figure is the operating range of the production of electricity and exhaust heat output when operating PEFC or NEG. Operation Area B is the operating range of the production of electricity and the heat output (the amount of exhaust heat, and the boiler output) when operating a boiler in addition to the operation of PEFC or NEG. However, the boiler efficiency of Operation Area B was set at 0.9. In the operation map, the city gas calorific power (described as the amount of city gas consumption below) supplied to a system is described. When the operation map of PEFC (Figure 5.6(a)) and NEG (Figure 5.6(b)) is compared, IEG of Operation Area A is larger. Furthermore, in Operation Area A of IEG, if the production of electricity increases, the exhaust heat output will also increase, but in IPF, even if the production of electricity increases, the increase in exhaust heat output is lost. PEFC is assumed in this chapter to have the characteristic that the rate of heat output to production of electricity is small when the production of electricity increases.

5.4.2 Operation Map of HCGS

Figure 5.6(c) shows the operation map of compound operation of 5-kW PEFC and 5-kW NEG introduced into HCGS. In HCGS, PEFC is operated corresponding to a base load, and NEG is operated corresponding to fluctuation load. A base load will be 5 kW, the high-load region of PEFC is always highly efficient, and it can be operated. However, when the load is less than 5 kW, PEFC follows this load and is operated. The maximum efficiency point of PEFC is near the rate of the maximum load, as shown in Figure 5.5(b). On the other hand, NEG operates corresponding to load fluctuation, and operations in a low-load region occur frequently. NEG has good power-generation efficiency and exhaust gas characteristic in low load, as shown in Figure 5.4. Moreover, as for low-load operation of NEG, an effect is expected if the characteristic of the power-generation efficiency and the load factor of PEFC and NEG is the same as the model shown in Figure 5.1(c). Therefore, it is thought that power-generation efficiency improves HCGS compared with IPF and IEG. HCGS proposed in this chapter assumes the introduction to a demand model with much fluctuation of power load.

When the range of Figures 5.6(a) and (b) is compared, when the power-generation range is 2 to 10 kW, and the heat output range is 0 to 50 kW, the amount of city gas consumption of IEG is 11 to 73 kW, and it is 17 to 69 kW in IPF. Therefore, the IEG change in the amount of city gas consumption at the time of load fluctuation is larger than IPF. The difference in the amount of city gas consumption in Operation Area A of each system is greater. For example, in the power-generation range of 2 to 10 kW, the amount of city gas consumption of IEG is 11 to 44 kW. The amount of city gas consumption of IPF is 17 to 31 kW, and is from 10.1 to 43.2 kW in HCGS as shown in Figure 5.6(c). Therefore, even if the same load fluctuation as each system is added, IEG and HCGS require a rapid response from the reformer.

5.5 Case Study

5.5.1 Power and Heat Demand Model

Figure 5.7 shows the power and the heat-demand model of 10-household apartments in Sapporo, Japan used in the analysis, and this is an average load of each sampling time.[12] However, an actual power-demand pattern is a set of loads that change rapidly in a short time, such as an inrush current. The

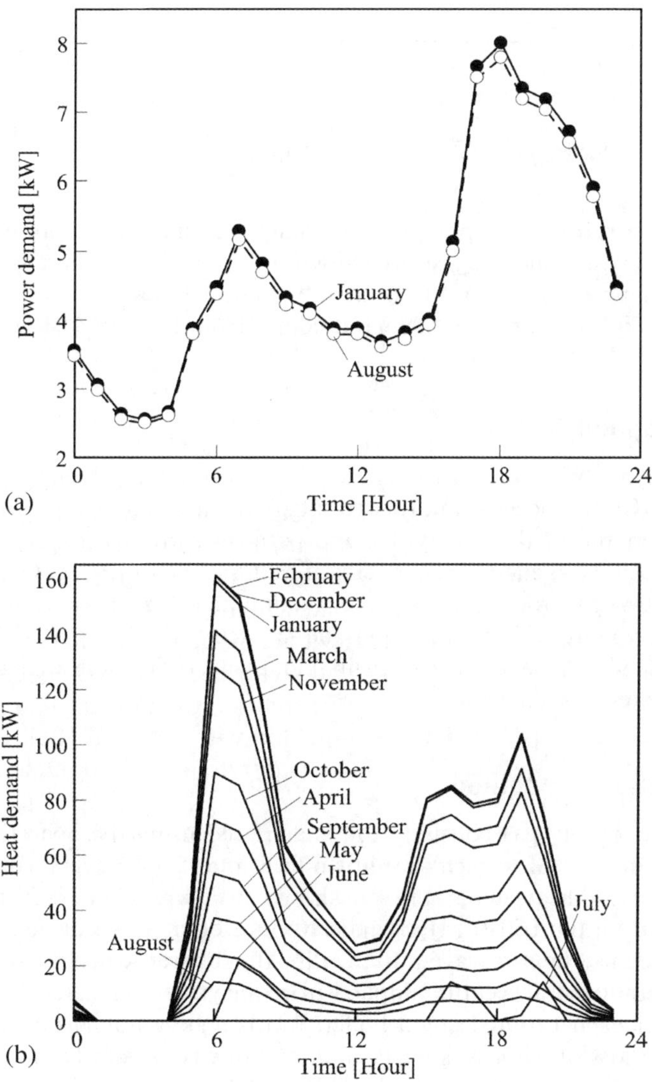

Figure 5.7 Power and heat-demand model for ten-household apartments in Sapporo; (a) Power-demand model; (b) Heat-demand model.

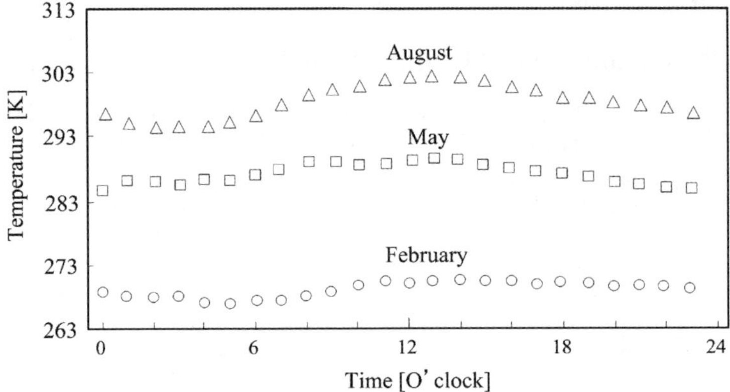

Figure 5.8 External temperature model in Sapporo.

average temperatures in Sapporo for the sampling times on a representative day in February, May, and August are shown in Figure 5.8.[13] There is no cooling load in summer in Sapporo. Electricity demand includes household appliances and electric lighting, and heat demand comes from heating, hot-water supply, and baths.

5.5.2 Capacity Setup

The capacity of NEG and PEFC in the case of introducing each system of IPF, IEG, and HCGS is decided in order to satisfy the maximum value of the power and heat demand of the energy-demand patterns shown in Figure 5.7. In this chapter, the power-generation capacity of each system of IPF, IEG, and HCGS will be 10 kW, in order to satisfy the maximum value of 8 kW of the power demand of Figure 5.7(a). Although the capacity distribution of PEFC and NEG of HCGS can be set up arbitrarily, it may be 5 kW in the analysis example of this chapter, respectively.

5.5.3 Analysis Method

The amount of city gas consumption is acquired using the operation map of IPF, IEG, and HCGS, which are shown in Figure 5.6 for every sampling time of the power and heat-demand model shown in Figure 5.7. However, when the exhaust heat output of IPF, IEG, and HCGS exceeds the heat demand, excess heat is stored in a heat-storage tank. On the other hand, when the output is less than the heat demand, heat is supplied from a heat-storage tank. When heat is still insufficient, it is output by a boiler. In the analysis of this chapter, radiation-of-heat loss of the heat-storage tank of a day is fixed at 10%. The amount of city gas consumption and the amount of thermal storage are calculated regarding all the sampling times on a representative day every month. The power-generation efficiency for every sampling time is obtained by dividing the power demand by the city gas consumption calorific power of the system. The value that divides "the value adding the production of electricity and the amount of heat output" by "the city gas consumption calorific power of the system" is defined as overall efficiency. A load factor can be given to Figure 5.4 and Figure 5.5(b), and the carbon-dioxide emissions of IPF, IEG, and HCGS can be obtained. The total discharge on a representative day is calculated by adding the amount of carbon-dioxide discharged by the boiler.

5.6 Results and Discussion

5.6.1 Operation Plan of a Representative Day

5.6.1.1 Amount of City Gas Consumption

Figure 5.9 shows the analysis results of each system for every sampling time on representative days in January and August. As Figure 5.7(a) shows, the power-demand pattern in January and August is almost the same. Therefore, there is no monthly difference with the high fuel consumption characteristic of each power generator (IPF, IEG, HCGS) (Figures 5.9(1)(a), and 5.9(2)(a)). However, the profiles of fuel consumption differ according to each power generator. The fuel consumption of NEG on a representative day shows large changes compared with other power generators. Moreover, HCGS shows least fuel consumption compared with other power generators. The difference in the profiles of fuel consumption is based on the operation map shown in Figure 5.6.

5.6.1.2 Heat System

Figure 5.9(1)(b), and Figure 5.9(2)(b) show the analysis results of the exhaust heat output of IPF, IEG, HCGS in January and August. There is a large change in the exhaust heat output of NEG compared with other power generators. Moreover, HCGS produces the least exhaust heat. The reason for HCGS producing the least exhaust heat is explained below.

(1) Figure 5.7(a) shows the power-demand pattern used in this chapter, which is 5 kW or less. Therefore, the exhaust heat of HCGS of Figure 5.6(c) is output by a fuel cell in many cases. Even if the production of electricity increases with PEFC, the increase in exhaust heat output is lost. Furthermore, the capacity of PEFC is 5 kW and is small compared with IPF (10 kW). Therefore, there is little exhaust heat output of HCGS compared with IPF and IEG.
(2) HCGS has the least fuel consumption, as Figures 5.9(1)(a), and 5.9(2)(a) show. Therefore, since the power-generation efficiency is high compared with other systems, HCGS has little exhaust heat output.

Figures 5.9(1)(c) and 5.9(2)(c) show the analysis results of the amount of thermal storage in January and August. Although heat is stored from midnight

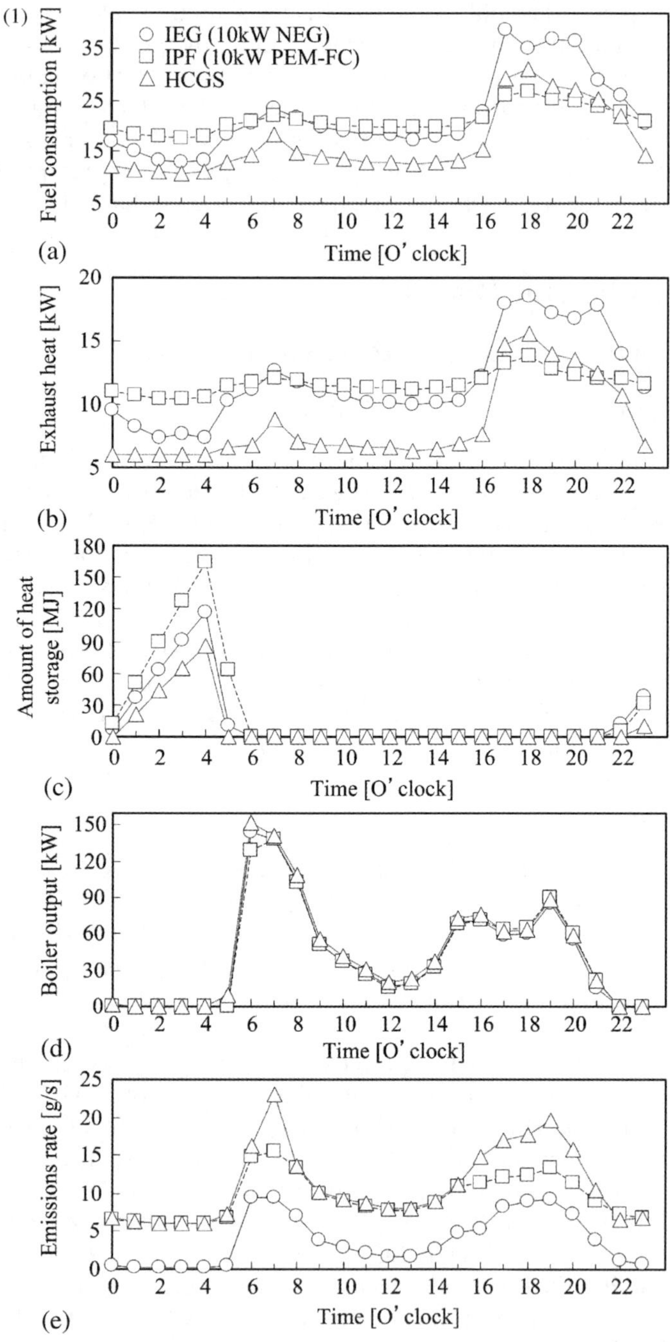

Figure 5.9 Analysis results of representation day; (1) January; (a) Fuel consumption of a power generator; (b) Exhaust heat output; (c) Amount of heat storage; (d) Boiler output; (e) Carbon-dioxide emissions rate. (2) August; (a) Fuel consumption of a power generator; (b) Exhaust heat output; (c) Amount of heat storage; (d) Carbon-dioxide emissions rate.

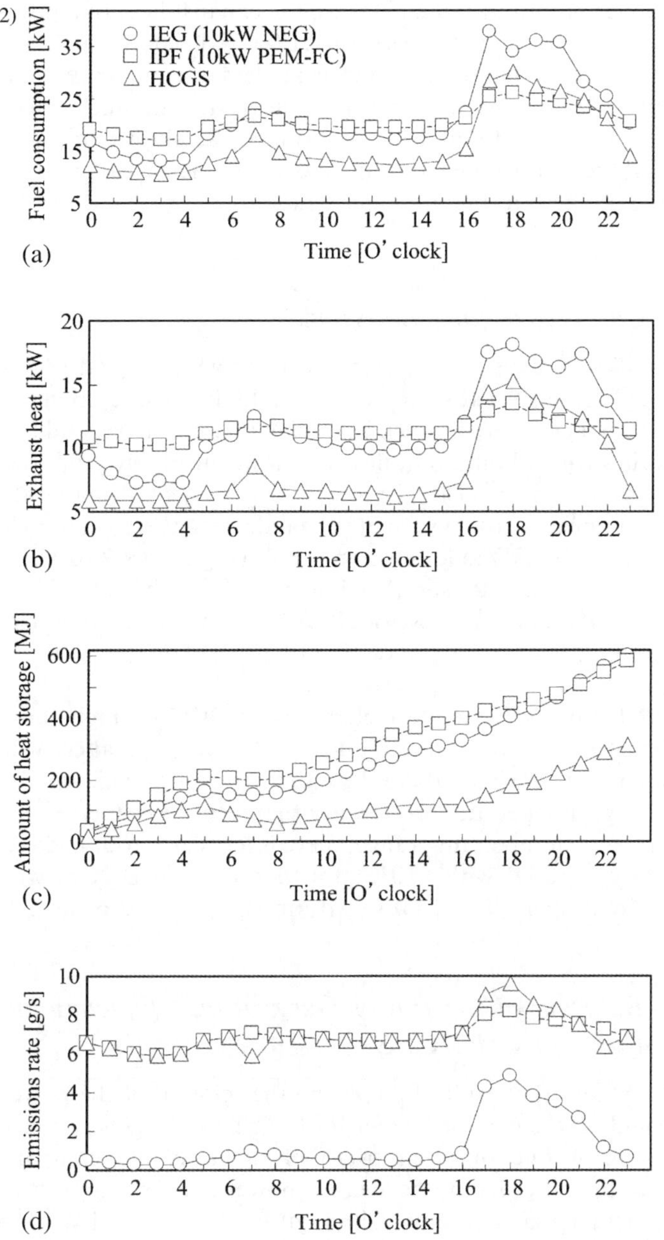

Figure 5.9 Continued.

to the early morning in January, heat is consumed in the morning. From 21:00 to the following morning (5:00 or 6:00), since the exhaust heat of a power generator is insufficient in January compared with the heat demand, insufficient heat is supplied from the boiler at that time. Figure 5.9(1)(d) shows the analysis results of boiler output in January. Since the exhaust heat of a power generator is less than the heat demand between 21:00 and 6:00, heat is supplied by a boiler. On the other hand, since the heat demand on representative days is fulfilled only by power-generator exhaust heat, the amount of thermal storage increases in August without being consumed, as shown in Figure 5.9(2)(c). Therefore, there is no boiler output in August. The heat of Figure 5.9(2)(c) needs to radiate heat to the air using a radiator.

5.6.1.3 Carbon-Dioxide Emissions

The carbon-dioxide emissions of each system on a representative day in January are shown in Figure 5.9(1)(e). The sources of discharge of carbon dioxide are the reformer and boiler. There is much heat demand in January, and there are many periods during which boiler output is as shown in Figure 5.9(1)(d). Therefore, the profile of Figure 5.9(1)(e) is similar to the heat demand characteristics described in Figure 5.7(b). On the other hand, since there is no boiler operation in August, the carbon-dioxide emission profile (Figure 5.9(2)(d)) is not similar to the heat demand characteristics. In January, the carbon-dioxide emissions of IEG are the minimum and those of HCGS, the most. The reasons for the high level of carbon-dioxide emission of HCGS are explained below.

(1) Since the power-generation efficiency of HCGS is high and there is little exhaust heat output, the frequency of boiler operation will increase in January. Therefore, there is more carbon-dioxide emission accompanying the operation of a boiler than other systems.
(2) In a low-load operating range, there is less carbon-dioxide emission with NEG compared with PEFC. Furthermore, in HCGS, we can see the high-load operation of NEG with much carbon-dioxide emission.

5.6.1.4 Relation of the Power-Generation Efficiency and CO_2 Emissions of HCGS

Figure 5.10 shows the result of analyzing the relation of the power-generation efficiency and CO_2 emissions of HCGS. The power-generation efficiency of about 5 kW of production of electricity of HCGS is high compared with IEG and IPF. On the other hand, the average power load of a representative day of the power-demand model shown in Figure 5.7(a) is 4.7 kW. Moreover, the power-generation efficiency of the low loading range of HCGS is clearly high compared with IPF. Therefore, the efficiency characteristic of HCGS shown in Figure 5.10 is suitable for installation to an apartment. The characteristic of CO_2 emissions will rise rapidly, if it exceeds 8 kW of production of electricity with few relative frequencies. The load characteristic shown in Figure 5.7(a) has few parts over 8 kW. In this way, the CO_2-emission characteristic of HCGS has the merit of IEG and IPF.

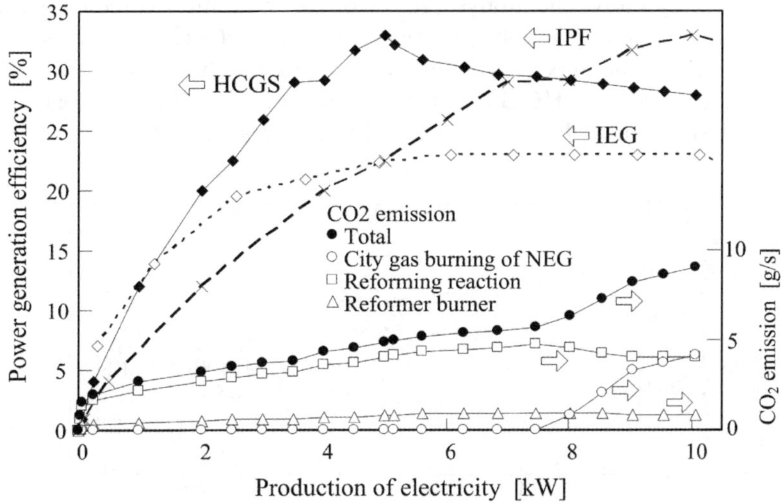

Figure 5.10 Relation of the power-generation efficiency and CO_2 emission of HCGS.

5.6.2 Annual Operation Plan

5.6.2.1 Annual Operation Plan of Heat System

Figure 5.11(a) shows the results of the amount of exhaust heat of the power generator on a representative day every month. The power-demand model (Figure 5.7(a)) used in this chapter has only a few differences each month. Since each power generator performs the following operation at power demand, the exhaust heat characteristics of each power generator do not have to be very different each month. On the other hand, the profile of the amount of exhaust heat changes with the type of system. The reason for HCGS having the least exhaust heat is the same as that explained in Section 5.6.1.2.

Figure 5.11(b) shows the results of the amount of exhaust heat output from each system every year. Although the difference in the amount of exhaust heat of IEG and IPF is small, the amount of exhaust heat has less HCGS than other systems. Since HCGS has little exhaust heat output, it is disadvantageous in cold-region winters with much heat demand. However, in the summer when exhaust heat remains, it is effective.

Figure 5.11(c) shows the analysis result of the boiler output of each system. Since HCGS has little exhaust heat output, there is a large amount of boiler output compared with other systems.

Figure 5.11(d) shows the result of the quantity of heat output by the boiler every year. The boiler output of HCGS is 1.12 times compared with other systems. This result suggests that HCGS should be introduced where there are conditions of little heat demand.

Figure 5.11(e) shows the analysis result of the maximum value of the boiler output of every month. We have to decide the boiler capacity of each system as a value exceeding the maximum value of each system shown in Figure 5.11(e). The boiler capacity of HCGS is 155 kW for most of the time. The boiler capacity of IEG and IPF is 147 kW and 140 kW, respectively.

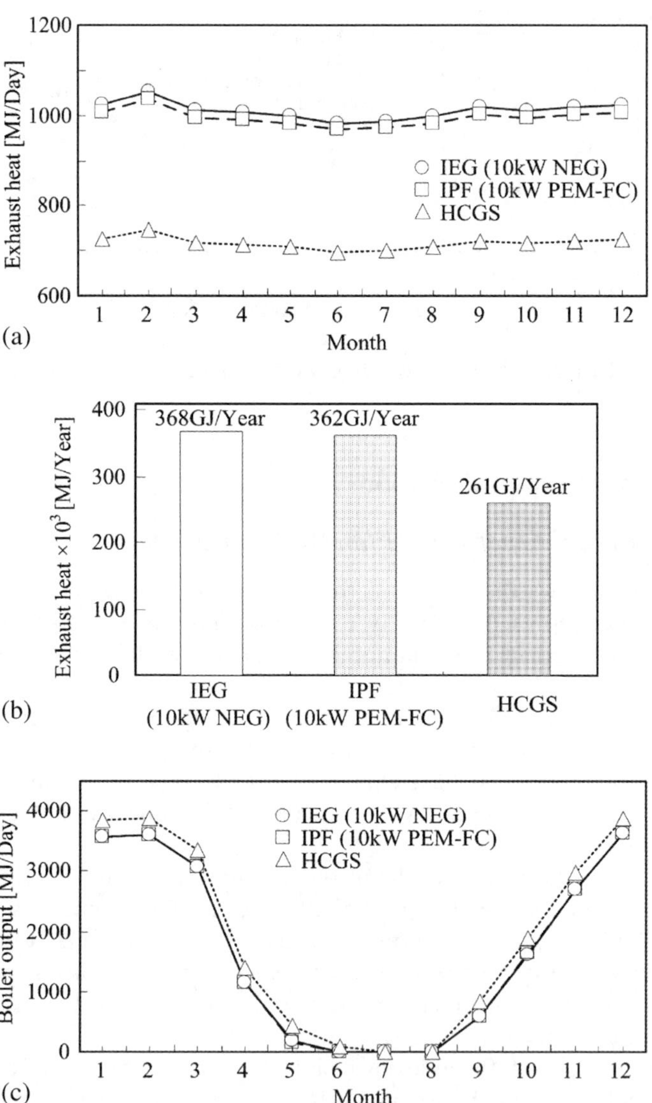

Figure 5.11 Analysis results of heat system and capacity; (a) System exhaust heat output; (b) Annual system exhaust heat output; (c) Amount of a boiler output; (d) Annual amount of a boiler output; (e) Boiler maximum output; (f) Amount of the maximum heat storage.

Figure 5.11(f) shows the analysis result of the monthly amount of the maximum thermal storage of each system. The capacity of the heat-storage tank introduced into each system needs to be decided so that it exceeds each maximum value shown in Figure 5.11(f). The heat-storage tank capacity of HCGS is the smallest at 100 MJ. The heating storage capacities of IEG and IPF are 135 MJ and 183 MJ, respectively. In IPF, the amount of

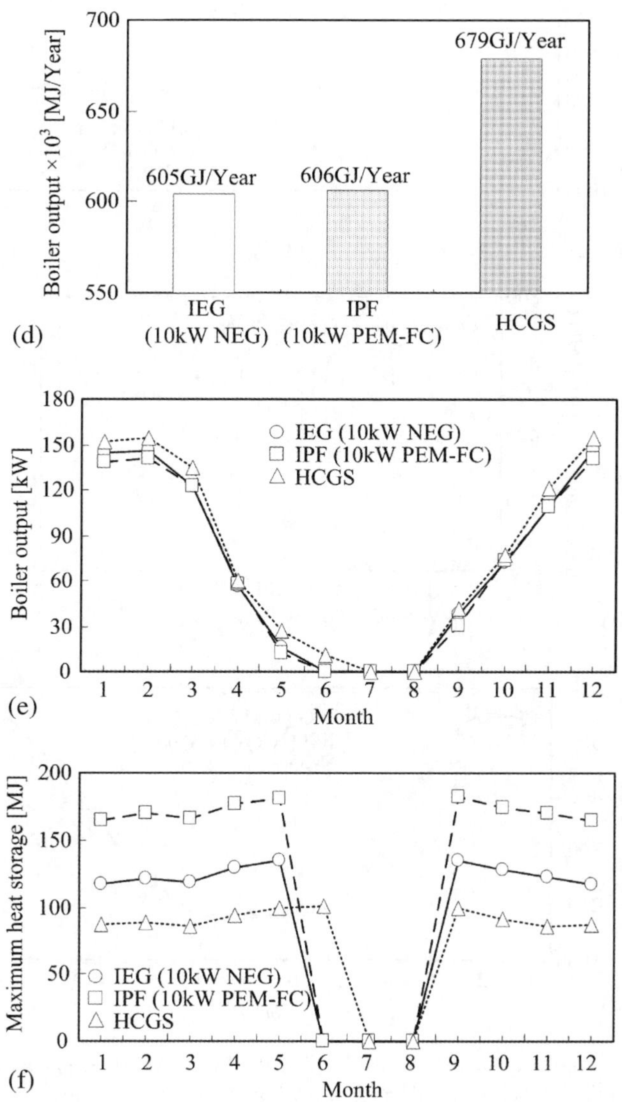

Figure 5.11 Continued.

change of the exhaust heat output is large for changing the production of electricity so that it can be found from an operation map. Therefore, in order to use all the exhaust heat of IPF, a heat-storage tank serves as a large-capacity tank.

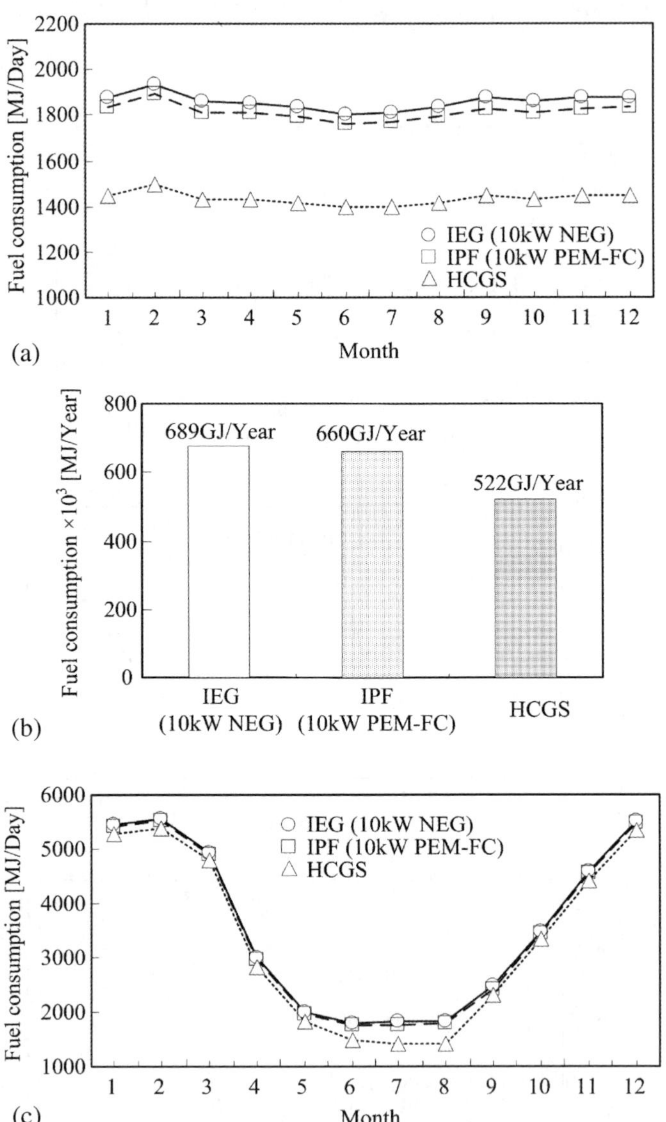

Figure 5.12 Analysis results of fuel consumption and carbon-dioxide emissions; (a) Fuel consumption of a power generator; (b) Annual fuel consumption of a power generator; (c) Fuel consumption of a power generator and a boiler; (d) Annual fuel consumption of a power generator and a boiler; (e) Carbon-dioxide emissions; (f) Annual carbon-dioxide emissions.

5.6.2.2 Fuel Consumption

Figure 5.12(a) shows the analysis results of the fuel consumption of each power generator (IEG, IPF, HCGS) on a representative day every month. The system with least fuel consumption every month is HCGS. This result shows that the power-generation efficiency of HCGS is higher than IPF and IEG. Figure 5.12(b) shows the annual amount of fuel consumption of each power

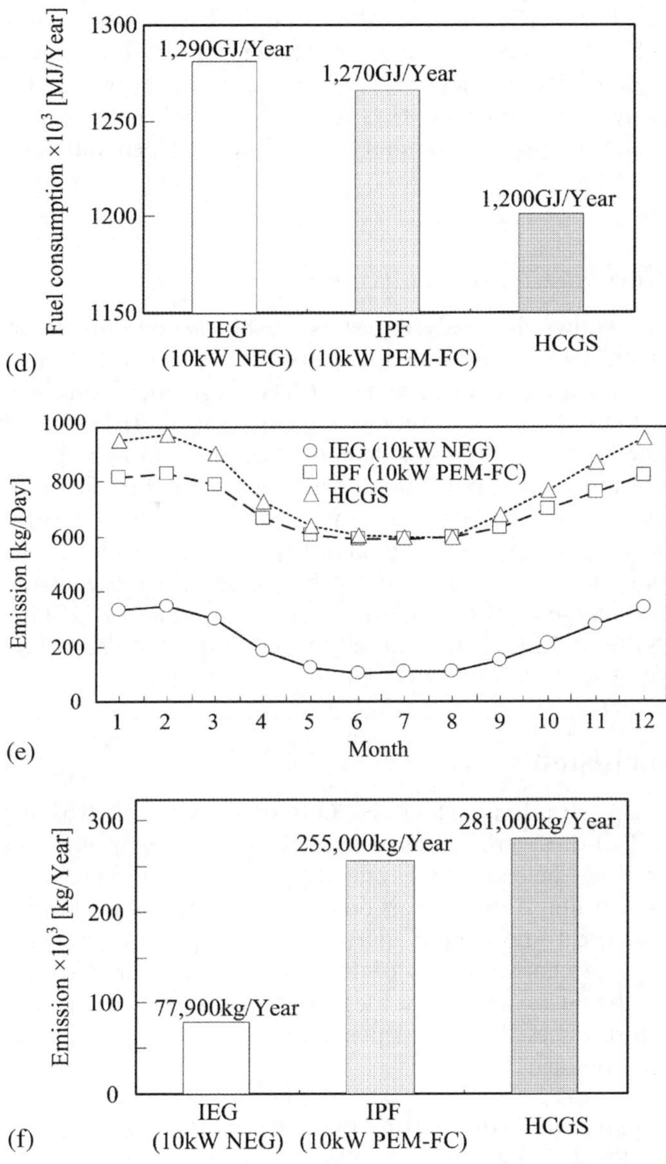

Figure 5.12 Continued.

generator. The annual fuel consumption of HCGS is 76% of IEG, and is 79% of IPF.

Figure 5.12(c) shows the analysis result of adding the fuel consumption of the boiler to the fuel consumption of each power generator shown in Figure 5.12(a). Since the energy-demand model used in this chapter has high heat demand, the result of Figure 5.12(c) is strongly influenced by the operation of the boiler. Figure 15.2(d) shows the results of adding the fuel consumption of an annual boiler to Figure 5.12(b). The annual fuel consumption of HCGS is 93% compared with IEG, and is 94% compared with IPF. Compared with Figure 5.12(b), there are few reduction effects of the fuel consumption of HCGS of Figure 5.12(d). Therefore, the conditions under which HCGS can be used effectively compared with IEG and IPF are when there is little boiler operation. Accordingly, a community with little heat demand can expect the introduction of HCGS.

5.6.2.3 Carbon-Dioxide Emissions

Figure 5.12(e) shows the analysis results of the carbon-dioxide emissions of each system on a representative day every month. Since the power-demand model used in the analysis of this chapter has many operation times in a low-load region, there are few carbon-dioxide emissions of IEG using NEG. The reason for the higher carbon-dioxide emissions of HCGS compared to IPF is that there are more operation times of a boiler than IPF. Therefore, the difference in the carbon-dioxide emissions between HCGS and IPF is large in winter when there is a high heat demand. Figure 5.12(f) shows the annual total of the carbon-dioxide emissions of each system. Compared with IEG, the amount of discharge of IPF is 3.3-fold higher, and that of HCGS is 3.6-fold higher. In NEG and PEFC introduced in this chapter, NEG is very advantageous regarding carbon-dioxide emissions.

5.7 Conclusion

The hybrid cogeneration system (HCGS) that uses a hydrogen-gas engine (NEG) and PEFC together was proposed concerning improvement of the power-generation efficiency at the time of partial load of fuel cell cogeneration. The city gas consumption and carbon-dioxide emissions in the case of introducing HCGS into 10-household apartments in Sapporo, Japan were analyzed. Furthermore, the characteristics of HCGS were compared with the system (IEG) that introduces an independent system (IPF) and the hydrogen-gas engine that introduces PEFC independently. As a result, the following conclusions were obtained.

(1) Compared with IPF or IEG, there is little fuel consumption of HCGS and the power-generation efficiency is higher. However, since the exhaust heat output is small in HCGS, if it is introduced into

conditions with high heat demand, the fuel consumption of the boiler will increase. Therefore, fuel consumption conditions with little heat demand and the power-generation efficiency of HCGS are more advantageous.

(2) IEG is the most advantageous regarding carbon-dioxide emissions. For example, concerning the power-demand pattern of a house, when operations in a low-load region occur frequently, IEG is especially advantageous. This is because a hydrogen-gas engine has good thermal efficiency in a low-load region compared with other systems.

(3) HCGS has more carbon-dioxide emissions than IPF and IEG. Since the power-generation efficiency of HCGS is good, there is little exhaust heat output. Therefore, if HCGS is introduced where the heat demand is high, the operation time of the boiler will increase. If the amount of carbon-dioxide discharged by the operation of the boiler and power generation is added, HCGS is disadvantageous compared with IPF or IEG.

(4) Operation planning and capacity planning of a boiler and a heat-storage tank were investigated. As a result, HCGS has the largest boiler capacity and IPF has the largest heat-storage tank capacity.

Acknowledgments

This work was partially supported by a Grant-in-Aid for Scientific Research(C) from JSPS.KAKENHI (17510078).

Nomenclature

E: power kW
g: CO_2 emission g/s
g': CO_2 emission g/(s kW)
$L_{c,rg}$: work of compression of a compressor N m
P: pressure Pa
q: calorific valve kW
t: sampling time
U: quantity of flow m^3/s

Greek Symbols

η: efficiency %

Subscripts

B: burner installed in a reformer
F: fuel cell
G: generator

I: engine combustion
N: gas engine generator
R: reformer
rg: reformed gas
S: boiler
∞ : atmosphere

Equipment

C/O: carbon-monoxide oxidation equipment
CGS: cogeneration
DC–AC: DC–AC converter
FC: fuel cell
G/T: generator
HCGS: hybrid cogeneration
I/T: inverter
NEG: gas engine generator
PEFC: solid polymer membrane-type fuel cell

References

1. S. Obara and K. Kudo, Study of a Small-Scale Fuel Cell Cogeneration System with Methanol Steam Reforming Considering Partial Load and Load Fluctuation, *Trans. ASME J. Energy Res. Technol.*, 2005, **127**(4), 265–271.
2. T. Ogawa *et al.*, Study on energy saving of the cogeneration for houses, the 8th report: The Survey and Fuel Cell Installation Feasibility Study of Power Load in an Apartment House, *2005 Proceedings of the Meeting of the Society of Heating, Air-Conditioning and Sanitary Engineers*, 2005, **3**, 2037–2040 (in Japanese).
3. Mohammadi Ali. *et al.*, Development of Highly Efficient and Clean Engine System using Natural-Gas and Hydrogen Mixture Fuel Obtained from Onboard Reforming, *NEDO report ID:03B71006c*, 2005 (in Japanese).
4. Yap D., *et al.*, Effect of Hydrogen Addition on Natural Gas HCCI Combustion, *SAE Tech. Pap. Ser.*, 2004, SAE-2004-01-1972.
5. J. F. Larsen and J. S. Wallace, Comparison of Emissions and Efficiency of a Turbocharged Lean-Burn Natural Gas and Hythane Fuelled Engine, *ASME ICE*, 1995, **24**, 31–40.
6. M. R. Swain and M. J. Usuf, The Effects of Hydrogen Addition on Natural Gas Engine Operation, *SAE Tech. Pap. Ser.*, 1993, SAE-932775.
7. T. Nakamura and M. Sei, Energy Related Technology. High-Efficiency Fuel Processor for Fuel Cell System, *Technical Report*, 2002, **77**, 4–9Matsushita Electric Works, Ltd (in Japanese).
8. K. Oda *et al.*, A Small-scale Reformer for Fuel Cell Application, *Sanyo Tech. Rev.*, 1999, **31**(2), 99–106 (in Japanese).

9. Takeda, Y. *et al.*, Development of Fuel Processor for Rapid Start-up, *Proc. 20th Energy System Economic and Environment Conference, Tokyo, January 29–30*, ed., K. Kimura, 2004, 343–344 (in Japanese).
10. Mikkola, M., *Experimental Studies on Polymer Electrolyte Membrane Fuel Cell Stacks*, Master's thesis submitted in partial fulfillment of the requirements for the degree of Master of Science in Technology, Helsinki University of Technology, 2001, pp. 58–79.
11. Ibaraki Prefecture Government Office of Education, Modeling of hydrogen energy system, *High school active science project research report*, Ibaraki, Japan, 2002 (in Japanese).
12. Narita, K., *The research on unused energy of the cold region city and utilization for the district heat and cooling*, Ph.D. thesis, Hokkaido University, 1996.
13. National Astronomical Observatory, Rika Nenpyo, *Chronological Scientific Tables CD-ROM*, 2003, Maruzen Co., Ltd., Tokyo, Japan.

CHAPTER 6

PEFC/Engine Generator Compound Energy System (2)– Power-Generation Efficiency of an Independent Microgrid Composed of Distributed Engine Generators

SHIN'YA OBARA

6.1 Introduction

Application of microgrid technology provides a backup power supply in an emergency, peak cut of an electric power plant, and effective use of exhaust heat.[1–3] In addition, an independent microgrid system (IMG: independent microgrid) that supplies electric power and heat without interconnecting with other power systems realizes the advantages of a distributed power source. It allows the construction of a power supply system with low environmental impact that uses renewable energy including green energy and unused energy. Furthermore, since the energy transport distance of a microgrid is short, it uses exhaust heat effectively. Therefore, overall efficiency is improved compared with conventional power-generation systems. However, IMG is required for the rapid dynamic characteristics that follow changes in power load compared with a grid interconnected for other systems.[4] Although the independent microgrid

RSC Energy Series No. 3
Compound Energy Systems: Optimal Operation Methods
By Shin'ya Obara and Arif Hepbasli

Published by the Royal Society of Chemistry, www.rsc.org

realizes maximum distributed energy, there are issues to tackle such as the stability of the dynamic characteristics of power and the development of an optimal design method.[5] In addition, potential drops in efficiency at partial load operation of the generating equipment occurring in the load fluctuations of a grid need to be improved. The change factor of the power load changes rapidly, such as during an inrush current, or a change in a long period for the demand. This chapter describes changes in a short period as load fluctuation, and changes in a long period as demand fluctuation.

When a building linked to a grid is a house, in IMG, both load fluctuation and demand fluctuation are expected to be large. In a large-scale power plant, since power is supplied to various demands, demand fluctuation is smoothed. In IMG with a big load fluctuation, if no electricity storage system is installed, the operating point of the power generator will change significantly. However, if facility cost and maintenance cost are taken into consideration, installation of an electricity storage system can be avoided. When not installing electricity storage equipment in IMG, a power generator with partial-load operations is expected to frequently operate at low efficiency, and the power-generation efficiency in total is expected to be lower than that in existing large-scale electric power facilities. In this chapter, the power-generation efficiency when constituting IMG from a distributed small diesel-engine power generator is investigated by numerical analysis. The aim is to control the number of operations of the distributed power generator, and raise the load per power generator and decrease operation with a partial load, which has low efficiency. The fall in the power-generation efficiency is due to partial load and the transportation loss of exhaust heat decreases, so that the capacity of the power generator is reduced and the number of distributions is increased. However, if the number of power generators increases, the facility cost and the maintenance cost rise. Therefore, in this chapter, the relationship between the number of diesel-engine power generators for distribution, power-generation efficiency and power cost is investigated by numerical analysis.

6.2 System Description

6.2.1 Independent Microgrid Configuration

The schematic figure of IMG constituted from the engine generator assumed in this chapter is shown in Figure 6.1. An engine generator can be installed in arbitrary buildings. An engine generator consists of a diesel engine and a synchronous power generator. The power generated with the engine generator is supplied to each building through a system-interconnection device and IMG. An engine generator can be installed in the machinery room of another building besides the machinery room of each building. A system-interconnection device also allows the connection of the load and IMG. The synchronization of the generated alternating current power and the power of IMG is controlled by a system-interconnection device.

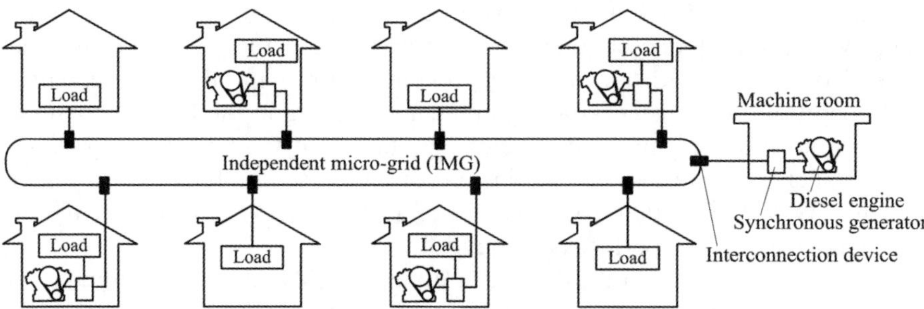

Figure 6.1 Diesel-engine-independent microgrid system.

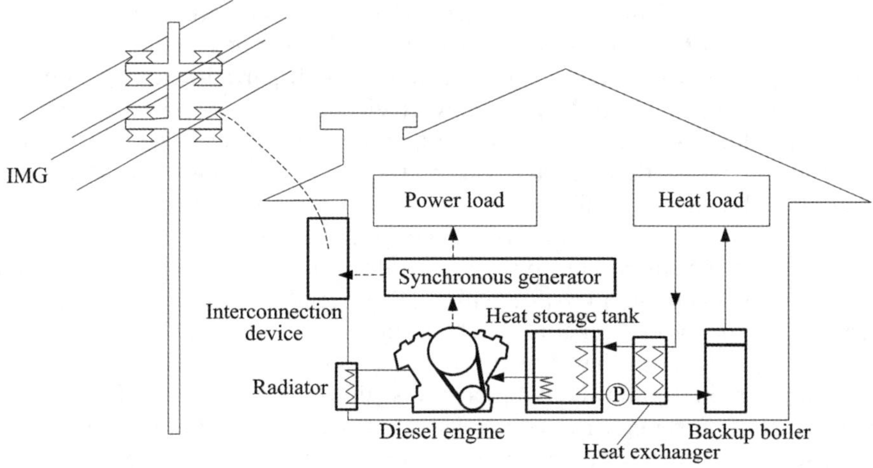

Figure 6.2 Diesel engine cogeneration system.

The system configuration of the engine generator installed in each building is shown in Figure 6.2. The exhaust heat output with an engine generator is engine-cooling water and exhaust gas, and this heat is stored in a heat-storage tank through a heat exchanger. In addition, the heat-transfer-medium heat exchanger for supplying heat to a building is installed in the heat-storage tank. After exchanging this heat-transfer medium for exhaust heat in the heat-storage tank, it is supplied to a backup boiler. When there is a large amount of exhaust heat compared with the heat load, the excess exhaust heat is stored in the heat-storage tank. On the other hand, when the exhaust heat exceeds the heating storage capacity of the heat-storage tank, the exhaust heat is released from a radiator.

6.2.2 Control of the Number of Engine Generators

Figure 6.3(a) shows the power-generation efficiency model of the central system of an engine generator, and Figure 6.3(b) shows the power-generation efficiency

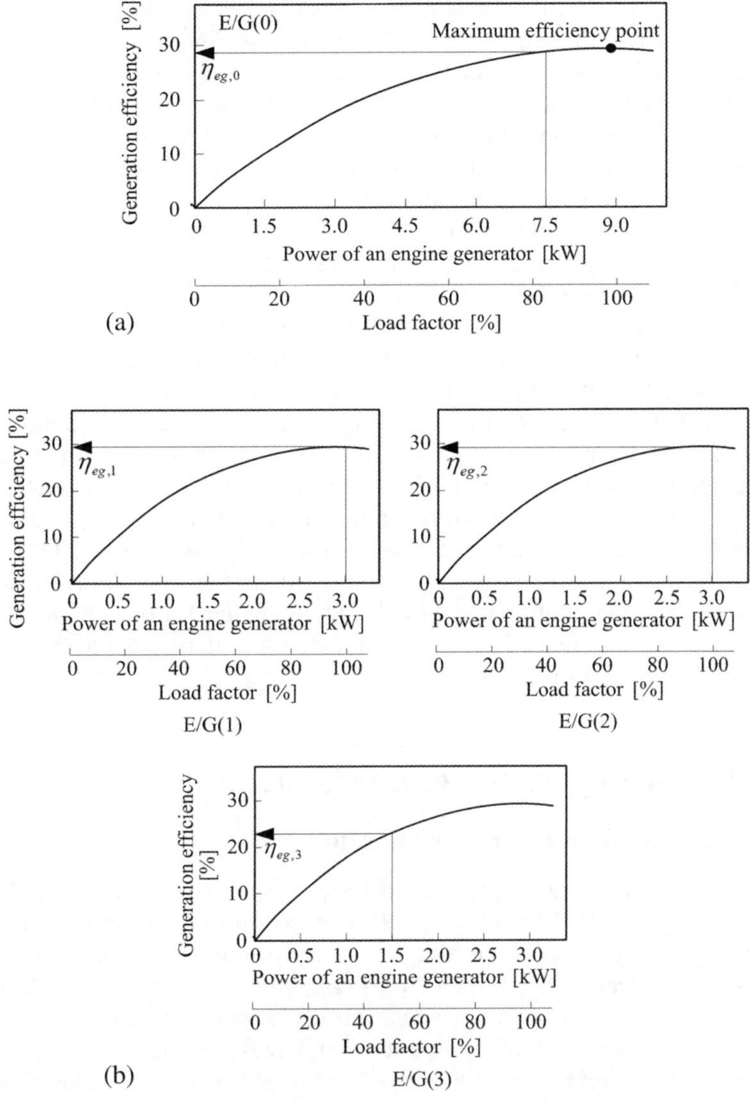

Figure 6.3 Installed numbers of engine generators, and efficiency model; (a) Efficiency model of a central system; (b) Efficiency model when installing three engine generators.

model of distribution. The model of Figure 6.3(a) expresses the relation between the power of an engine generator with a power-generation capacity of 9 kW, and the load factor and efficiency. Here, the ratio of production of electricity to maximum production of electricity is defined as the load factor. As shown in Figure 6.3(a), the power-generation efficiency in the case of a 7.5 kW load with one set (E/G(0)) of engine generators is 29%. The load factor of the engine generator at this time is 82%. The model of Figure 6.3(b) is a model of the power, the efficiency, and the load factor when installing three sets of engine generators (E/G(1), E/G(2), E/G(3)) with a power-generation capacity of 3 kW. The number of operations of a three-set engine generator is controlled, and it is operated so that the power load may be followed. When the same load (7.5 kW) as Figure 6.3(a) is added, E/G(1) and E/G(2) are operated with a full load with high efficiency (respectively, 3 kW), and E/G(3) corresponds to the remaining loads (1.5 kW). Since the load of E/G(3) is small, E/G(3) is the operation of a partial load with low efficiency. E/G(1) and E/G(2) of the load factor of an engine generator are 100%, and E/G(3) is 50%. If the power-generation efficiency of the centralized system of Figure 6.3(a) and the distributed system of Figure 6.3(b) are compared, the distributed system may have a high load factor per set, and the power-generation efficiency in total may become advantageous. Therefore, this chapter investigates the difference in the power-generation efficiency of the centralized system of IMG and the distributed system. The maximum efficiency of a 9 kW commercial engine generator is high, about 2% compared with a 3 kW engine generator.[6] And, although a 3- to 20 kW engine generator is installed in the example reported in the case study below, the difference of the maximum efficiency of an engine generator is about 3%.[6] In this chapter, the difference in the efficiency due to engine generator size is not taken into consideration.

6.3 Diesel Engine Generator System

6.3.1 Engine Generator Specifications

The test equipment of the cogeneration system using a small-scale diesel engine is shown in Figure 6.4. The analysis of this chapter uses the experimental results obtained by operation of the test equipment in Figure 6.4. The engine specification of the cogeneration system of Figure 6.4 is shown in Table 6.1. In addition, the specifications of a synchronous power generator are shown in Table 6.2. The fuel of the diesel engine is kerosene, and the engine has 2 cylinders with 4 cycles. The power generator is a single-phase synchronous type, and the power is transmitted from the power axis of the diesel engine.

6.3.2 Output Characteristics of a Small-Scale Diesel Engine Cogeneration System

The heating value of the engine-cooling water, the heating value of engine exhaust gas, and the production of electricity are shown in Figure 6.5. If the kerosene fuel supplied to the engine is increased, the production of electricity and the heating value of the exhaust gas increases, but the heating value of the engine-cooling water decreases. The kerosene supply heating value at the time of 3-kW maximum output is 9.8 kW. Figure 6.6 shows the examination results

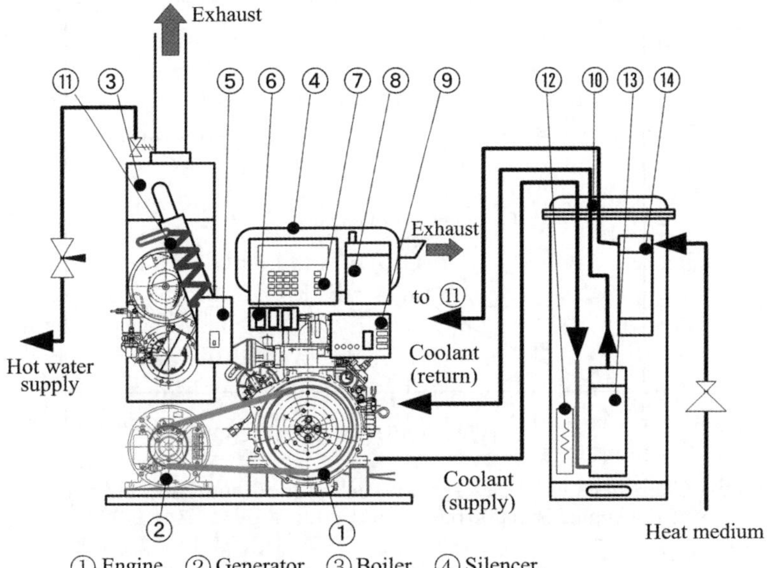

Figure 6.4 Test diesel engine cogeneration system.

Table 6.1 Engine specifications.

Engine type	Vertical straight 4 cycle diesel
Number of cylinders	2 cylinders
Total stroke volume	451 cc
Combustion type	Special swirl chamber
Compression ratio	24.5
Fuel	Kerosene
Size	369×385×485 mm
Dry weight	60 kgf

Table 6.2 Generator specifications.

Generator type	Single-phase synchronized
Rated output	5 kVA
Rated voltage	100 V
Rated electric current	50 A
Frequency	50 Hz
Number of revolution	3000 rpm
Size	200×221×359 mm
	Automatic voltage regulator

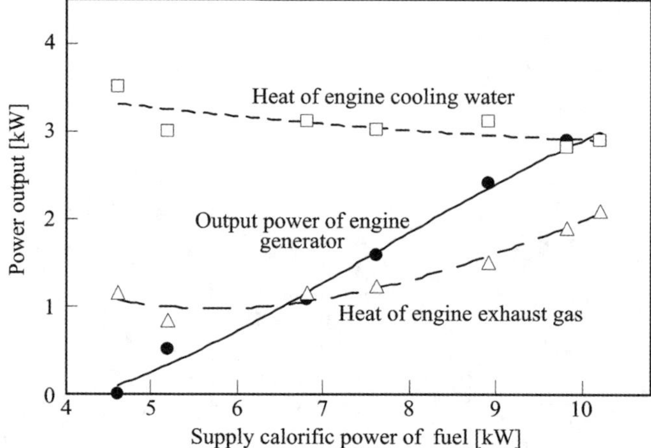

Figure 6.5 Characteristics of diesel engine generator output. The number of revolutions of the engine is 1600 rpm.

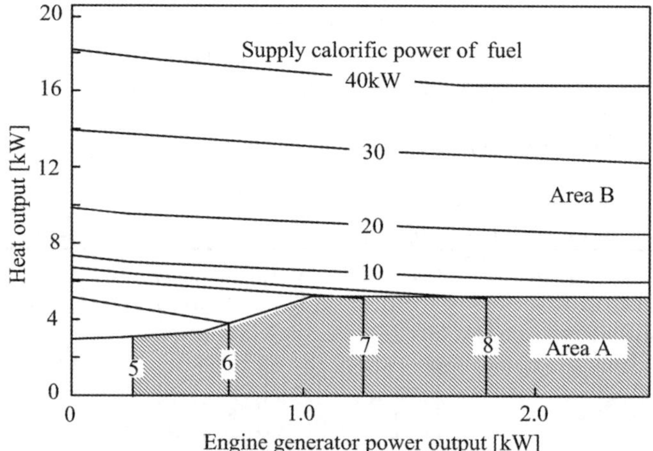

Figure 6.6 Fuel supply and energy output of engine generator. The number of revolutions of the engine is 1600 rpm. Area A corresponds to the condition in which only the engine-generator is working. Area B corresponds to the condition where the boiler is added to the engine-generator.

of the relation between the amount of kerosene fuel supplied to the engine cogeneration shown in Figure 6.4, and the power and the exhaust heat output power of the system. The hatched area in Figure 6.6 (Area A) is the area of the power where the exhaust heat is output only with the engine generator, and the other areas (Area B) are areas that include the heat output power in a kerosene boiler in addition to the output power of Area A. Figure 6.7 shows the

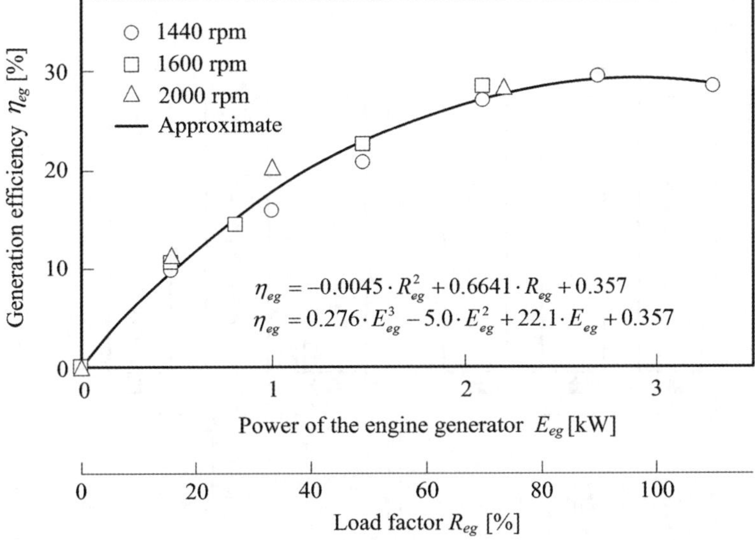

Figure 6.7 Diesel engine generator efficiency.

examination results of the production of electricity and the power-generation efficiency of a diesel-engine power generator, and the load factor and power-generation efficiency. Although the engine rotations were changed and examined from 1440 rpm to 2000 rpm, the influence on the power-generation efficiency was small. An approximate expression of each examination result is shown in the figure. These approximate expressions are used in the calculation of the power-generation efficiency of the case study. In an engine generator of the same form, even if the maximum engine production of electricity differs, the characteristics of load factor and power-generation efficiency can be managed similarly. Then, the relation between the load factor and the power-generation efficiency of the engine generator used in the case study is managed without being dependent on the power-generation capacity.

6.4 Case Study

6.4.1 Analysis Method

Figure 6.8 shows the analysis program of the model that distributes two sets of engine generators, E/G(1) and E/G(2). Each block in Figure 6.8 expresses a calculation component. Although E/G(1) is first operated in the analysis program, when the electricity demand exceeds the maximum production of E/G(1), E/G(2) is started. E/G(2) is started on the basis of the IF branch in Figure 6.8. Moreover, the production of electricity in all the engine generators in operation

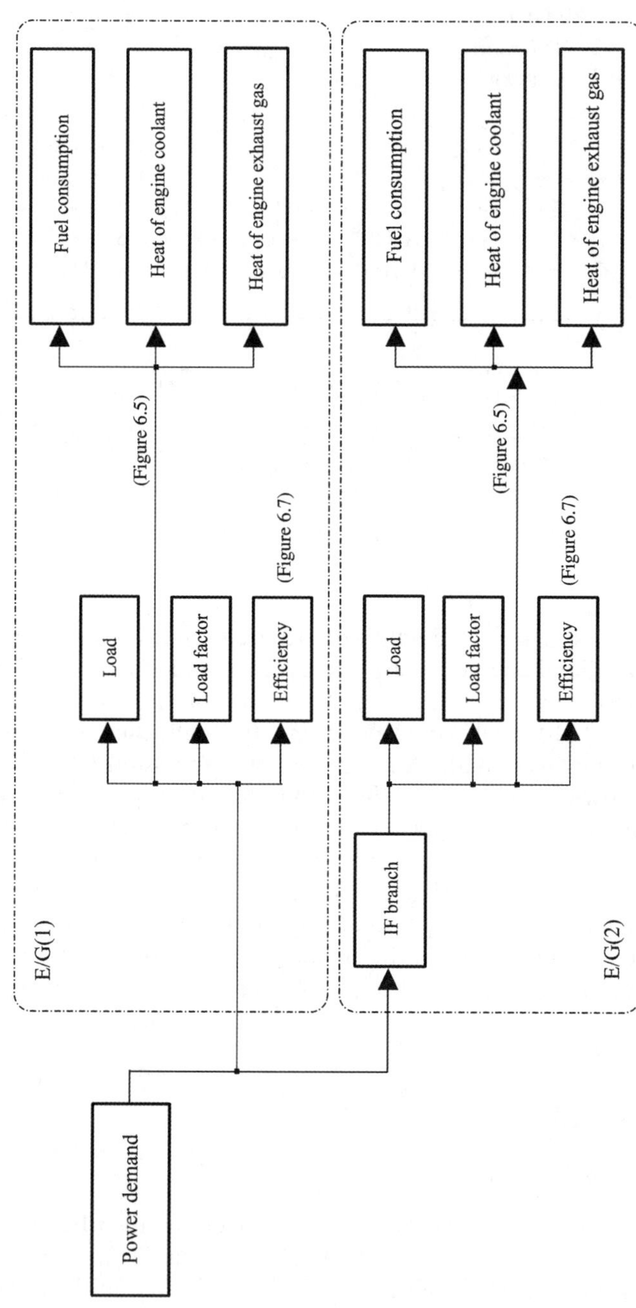

Figure 6.8 Block diagram of analysis. The number of diesel-engine generators to be distributed is two.

is equal to the electricity demand, and an engine generator that follows the power load is in operation. The power-generation capacity of each engine generator is input beforehand into the analysis program of Figure 6.8. Calculation of the fuel consumption, the heating value of the cooling water, and the heating value of the exhaust gas is performed using the engine power characteristics shown in Figure 6.5. If an electricity demand pattern is input into an analysis program and the number of operations of an engine generator is controlled, the power-generation efficiency can be obtained by introducing the load factor of each engine generator into Figure 6.7. The characteristics of the heat of the cooling water and the heat of the exhaust gas of each engine generator are calculated as being relatively the same as the experimental results of Figure 6.5, even if the power-generation capacity differs. In this chapter, the number of arrangements of an engine generator is analyzed regarding the model of 1 (central system: 18 kW, one set), 2 (9 kW, two sets), 4 (4.5 kW, four sets), and 6 (3 kW, six sets). The calculation of power cost takes the unit price of kerosene fuel. In this chapter, the kerosene cost in the U.S. is set at 409 $/m^3$.[7] Moreover, the conversion rate of the dollar to the yen was set at 1 dollar = 118 yen. The input–output characteristics of the power of IMG are analyzed using MATLAB (Ver. 7.0) and Simulink (Ver. 6.0) of Math Work Co. Ltd. Analysis error was managed within 0.01%.

6.4.2 Weather Conditions in Sapporo

The average temperatures in Sapporo for the sampling time on representative days in February, May, and August are shown in Figure 6.9.[8] There is no cooling load in the summer in Sapporo. Electricity demand includes household appliances and electric lighting, and heat demand comes from heating, hot-water supply, and baths. The area of an average individual house (a 3- or

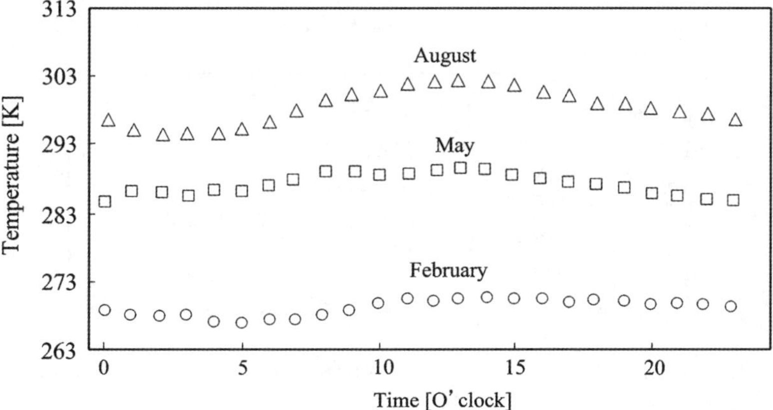

Figure 6.9 External temperature model in Sapporo.

4-person household) in Sapporo is 140 m^2 with 2 storeys, and the houses are made of wood.

6.4.3 Energy Demand Models

The power demand pattern of Figure 6.10(a) is the measurement result of an individual house for representative days in February in Sapporo, Japan.[9] Moreover, Figure 6.10(b) shows the heat-demand pattern of each month of an average house in Sapporo.[10] The electricity demand pattern does not change significantly each month. This is because there is no cooling load in the summer and the exhaust heat and the heat of the auxiliary boiler are supplied to a house assuming the heating load of winter. In the case study, the load factor, the heating value of the cooling water, the heating value of the exhaust gas, the fuel consumption, and the power-generation efficiency of each engine generator are calculated using the electricity demand pattern of the representative day shown in Figure 6.10(a). The analysis model of IMG connects and constitutes 20 houses of the electricity demand pattern of Figure 6.10(a).

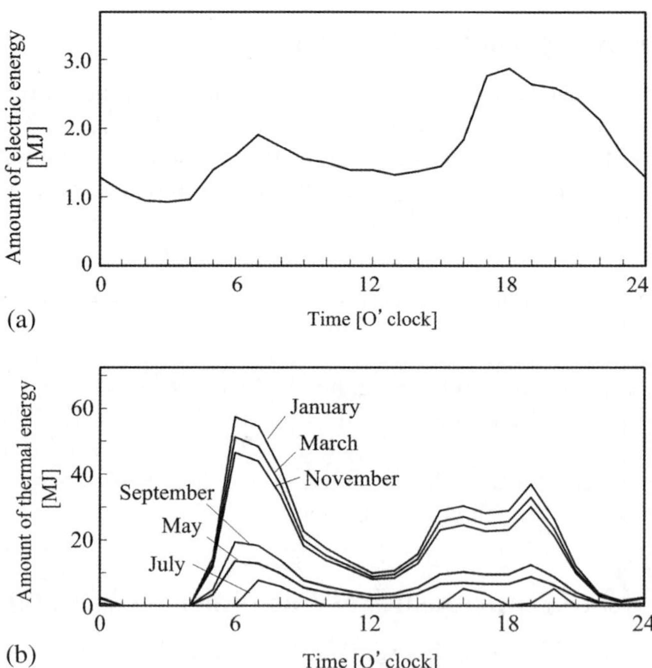

Figure 6.10 Average daily power and heat needs of a cold-region house with traditional heating system space; (a) Power needs; (b) Heat needs.

6.5 Results and Discussion

6.5.1 Load Distribution of the Engine Generator

Figure 6.11 shows the characteristics of the load added to each engine generator when distributing and arranging an engine generator and controlling the number of operations. In the central system, one set of engine generators with a power-generation capacity of 18 kW corresponds to the load of all sampling times. On the other hand, in Figures 6.11(b)–(d), it corresponds to the load by controlling the number of operations of two sets, four sets, and six sets, respectively, of engine generators. As a result of analysis, the load of each E/G(1) of Figures 6.11(b)–(d) is the largest, and the load of E/G(2) is the next largest. For example, in Figure 6.11(c), E/G(1) is the maximum power at all sampling times, and the power of E/G(2) has an operating time that is not the maximum output in the time zone of midnight to 6 o'clock in the morning, or the time zone of 8:00 to 16:00. E/G(4) is operated only in the time zone of 16:00 to 21:00. Using the method of controlling the number of operations as in this case, E/G(1) shows the highest production of electricity, and this decreases in the order of E/G(2), E/G(3), and E/G(4).

6.5.2 Number of Distributions, and Full Force Power

Figure 6.12 shows the number of engine generators for distribution, as well as the results of the breakdown of the energy output at each sampling time with each engine generator. Compared with the central system, fuel consumption on a representative day decreases in total by the system shown in Figures 6.12(b)–(d). However, in the 7:00 to 9:00 time zone, and 17:00 to 21:00 time zone, there is little reduction in the fuel consumption of the system of Figures 6.12(b)–(d). The reason is that Figure 6.11(b) is E/G(2), Figure 6.11(c) is E/G(3), and Figure 6.11(d) is E/G(4) and E/G(6), operating in the 7:00 to 9:00 time zone, and the 17:00 to 21:00 time zone, with a partial load with low efficiency as shown in Figures 6.11(b)–(d). Although E/G(5) of Figure 6.11(d) is operated in the 17:00 to 22:00 time zone, the load factor of E/G(5) at this time is 50% to 100%. Moreover, there is no operation in the other time zones. The 7:00 to 9:00 time zone and the 17:00 to 21:00 time zone show much higher fuel consumption compared with the time zones before and after each figure. The fuel consumption of an engine generator decreases in the order of Figures 6.12(a)–(d). Therefore, the fuel consumption resulted in a decrease, so that the number of distributions of the engine generator increased.

6.5.3 Output Characteristics of Each Engine Generator

Figure 6.13 shows the analysis results of the heat of the engine-cooling water, the heat of the engine exhaust gas, and the power generation of each engine generator for distribution. The heat of the engine-cooling water changes little, even if it changes the power load sharply according to the output characteristics

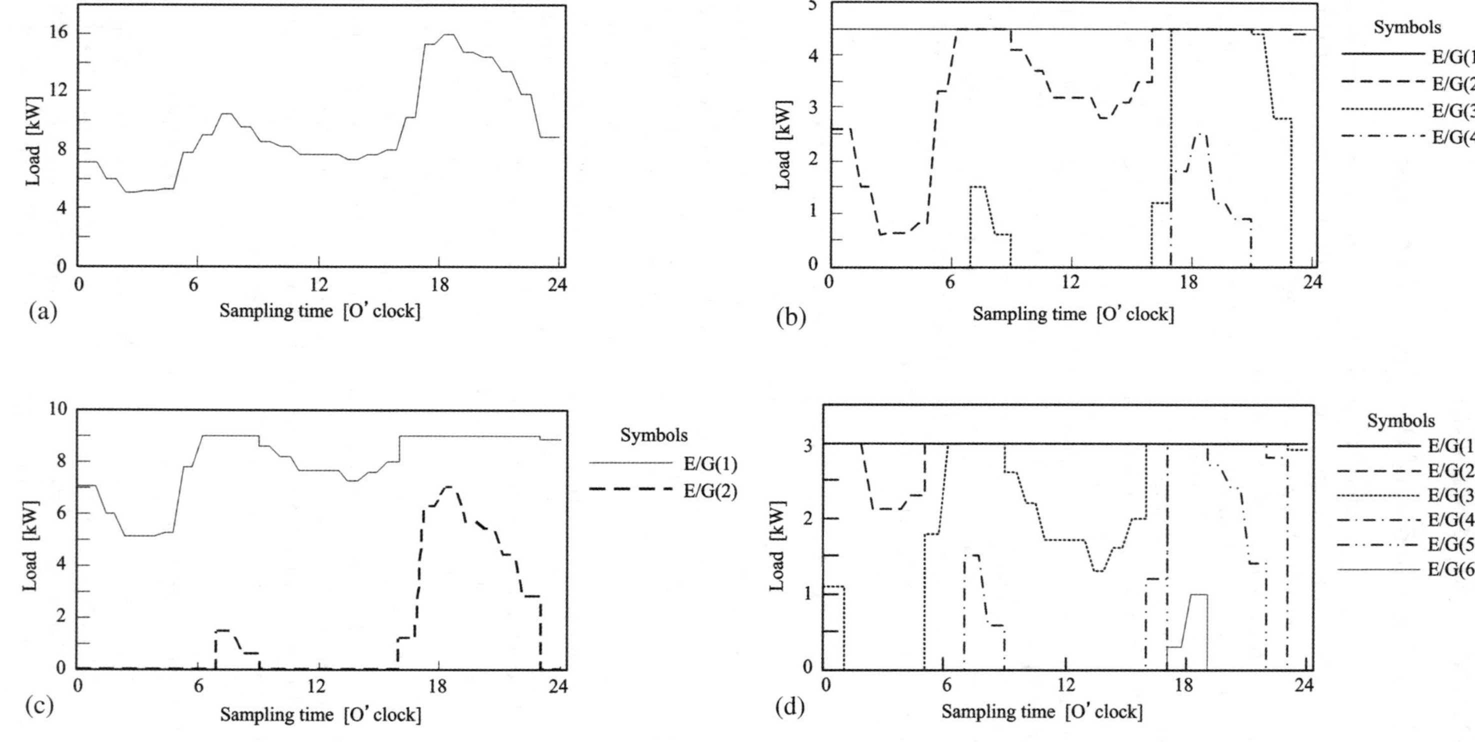

Figure 6.11 Load of each engine generator; (a) Central system; (b) Two systems; (c) Four systems; (d) Six systems.

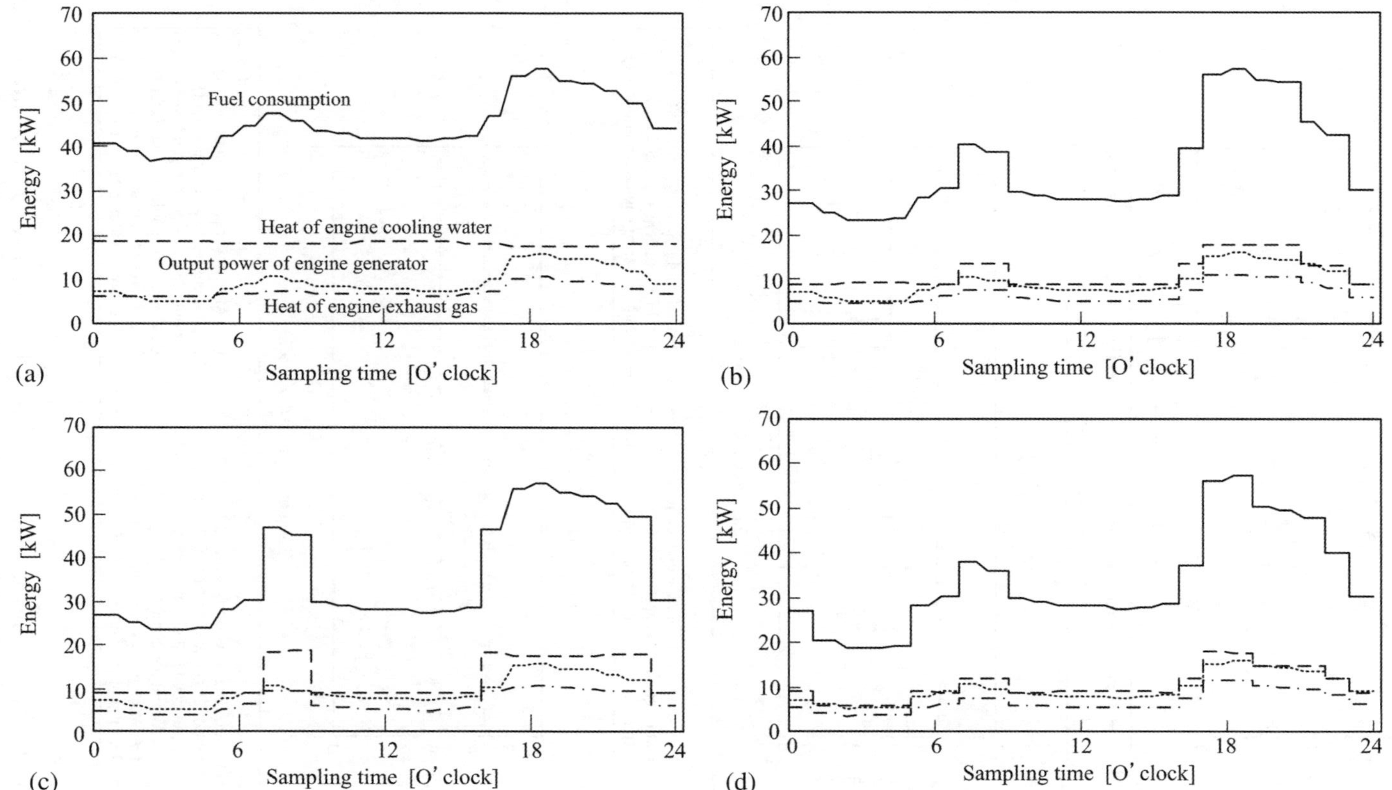

Figure 6.12 Installed numbers of engine generators, and total input-output of power; (a) Central system; (b) Two systems; (c) Four systems; (d) Six systems.

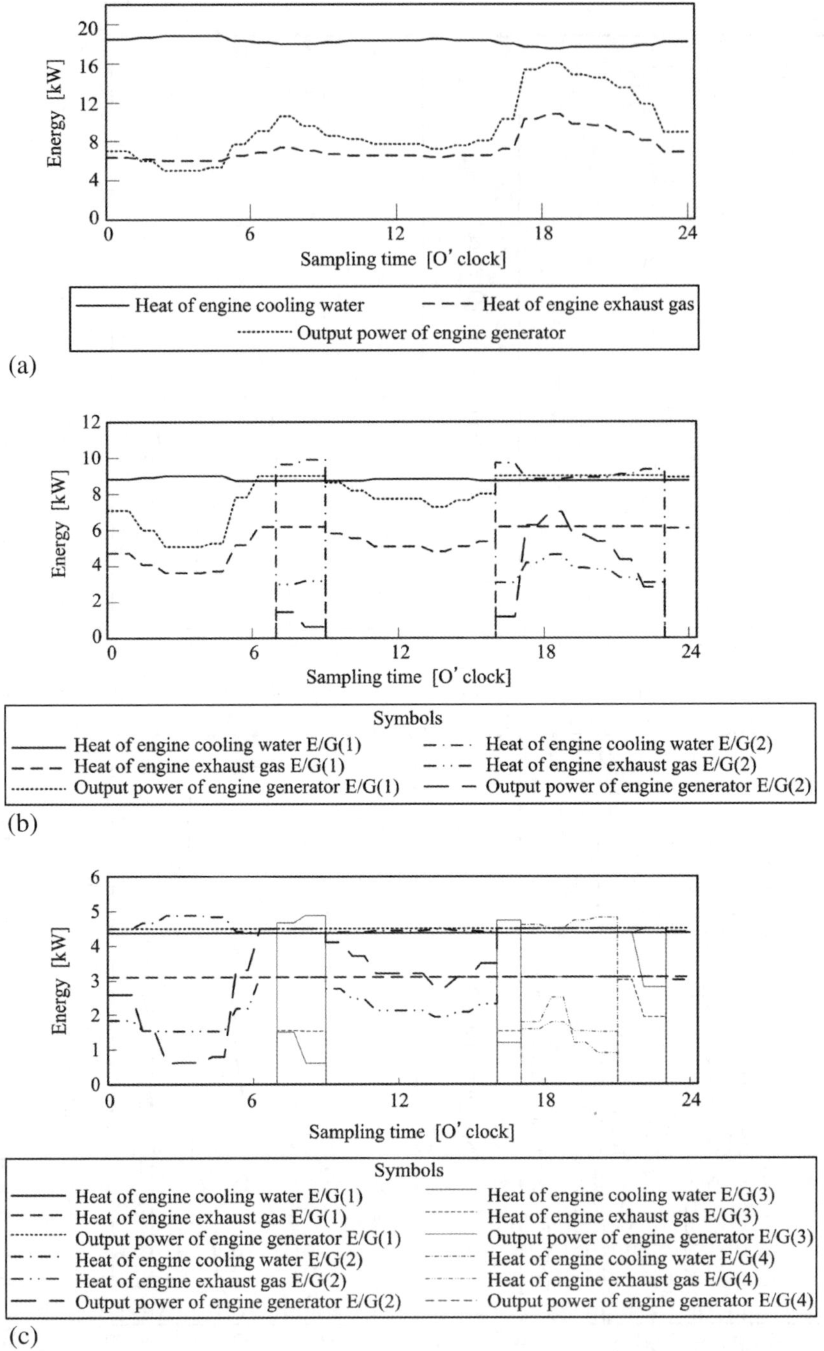

Figure 6.13 Analysis results of energy output; (a) Central system; (b) Two systems; (c) Four systems; (d-1) Six systems (1); (d-2) Six systems (2).

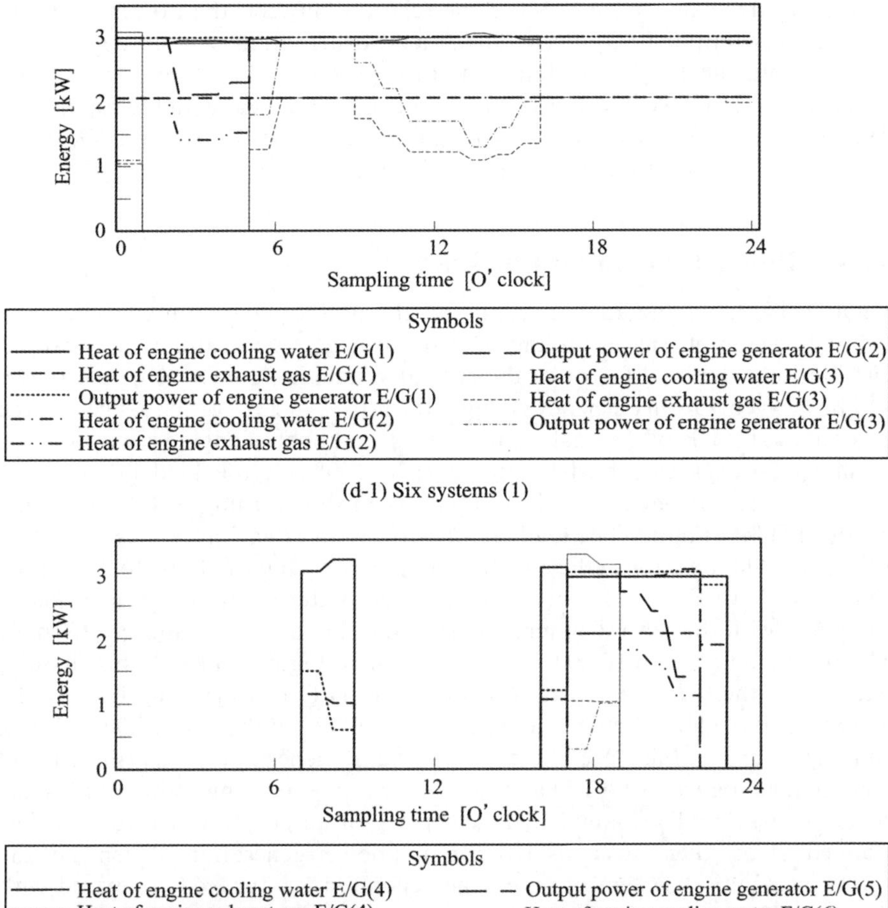

(d-1) Six systems (1)

(d-2) Six systems (2)

Figure 6.13 Continued.

of the engine generator shown in Figure 6.5. In the 7:00 to 9:00 time zone of Figure 6.13(b), the power generation of E/G(2) is 2 kW or less, and the load factor is 22% or less. As a result, in the 7:00 to 9:00 time zone of Figure 6.13(b), the heat of the engine-cooling water shows a large value as the output characteristics of Figure 6.5 show. Therefore, if the demand characteristics of the power and the heat in Sapporo shown in Figure 6.10 are taken into consideration and the engine generator operates from 23:00 to 5:00 with little electricity demand, a significant amount of heat from the engine-cooling water can be obtained. It can prepare for the large heat demand at 6 o'clock in the morning in winter (January, March, and November) shown in Figure 6.10(b) by storing this exhaust heat. Since the relation between the power load and exhaust heat output is nonlinear, if the load distribution of each engine generator is controlled, an operation plan fit for the demand pattern of power and heat may be realizable. However, this chapter does not deal with the supply and demand of heat. As for Figure 6.13(d–1), in the results of E/G(1) to E/G(3) of 6 system, (d-2) is the results of E/G(4) to E/G(6).

6.5.4 Power-Generation Efficiency

Figure 6.14 shows the analysis results of the power-generation efficiency and the load factor of each sampling time of an engine generator with arranged distribution. The load factor of the central system is 28 to 89% of the range, and power-generation efficiency changes from 13% in 28% of the range. On the other hand, E/G(1) of the analysis result of the number of engine generators is 6 as shown in Figures 6.14(d-1) and (d-2), is 100% of the load factor at all sampling times. From 2 o'clock to 5 o'clock in the morning, E/G(2) is a load factor of 70% or more. The load factor is 100% in other time zones. E/G(3) is 100% of the load factor in the 6:00 to 9:00 time zone and the 16:00 to 23:00 time zone. E/G(1) to E/G(3) all show an operation pattern with a high load factor compared with the central system. On the other hand, E/G(4) and E/G(5) have a high load factor from 16:00 to night as shown in Figure 6.14(d-2), but in other time zones, the load factor is a low-operation pattern. Regarding E/G(6), the operation of the load factor at a maximum of 33% is only from 16:00 to 22:00, and the hours worked are short. If an engine generator is distributed and arranged, as Figures 6.14(d-1) and (d-2) show, the operating ratio of an engine generator will be distributed. If the efficiency of an engine generator that has stopped is not taken into consideration, the power-generation efficiency of the whole distribution system of an engine generator is improved compared with the central system. However, the engine generator stops, and an engine generator with short working hours appears. If the life of the whole engine generator linked to IMG is taken into consideration, periodically replacing the starting order of the engine generator of E/G(1) to E/G(6) in Figures 6.14(d-1) and (d-2) is required, for example.

6.5.5 Power Cost

Figure 6.15 shows the analysis results of expressing the number of distributions of an engine generator, and the consumption of fuel for power generation on a representative day with the calorific power of fuel. If the number of distributions of an engine generator increases as shown in the figure, the fuel consumption will decrease. However, when four distributions of an engine generator are compared with six, the difference is 6%, and even if the number of installations is further increased, the reduction effect of fuel consumption shows slight improvement. If the facility cost and maintenance cost of an engine generator can be estimated, the number of years required to recover costs is

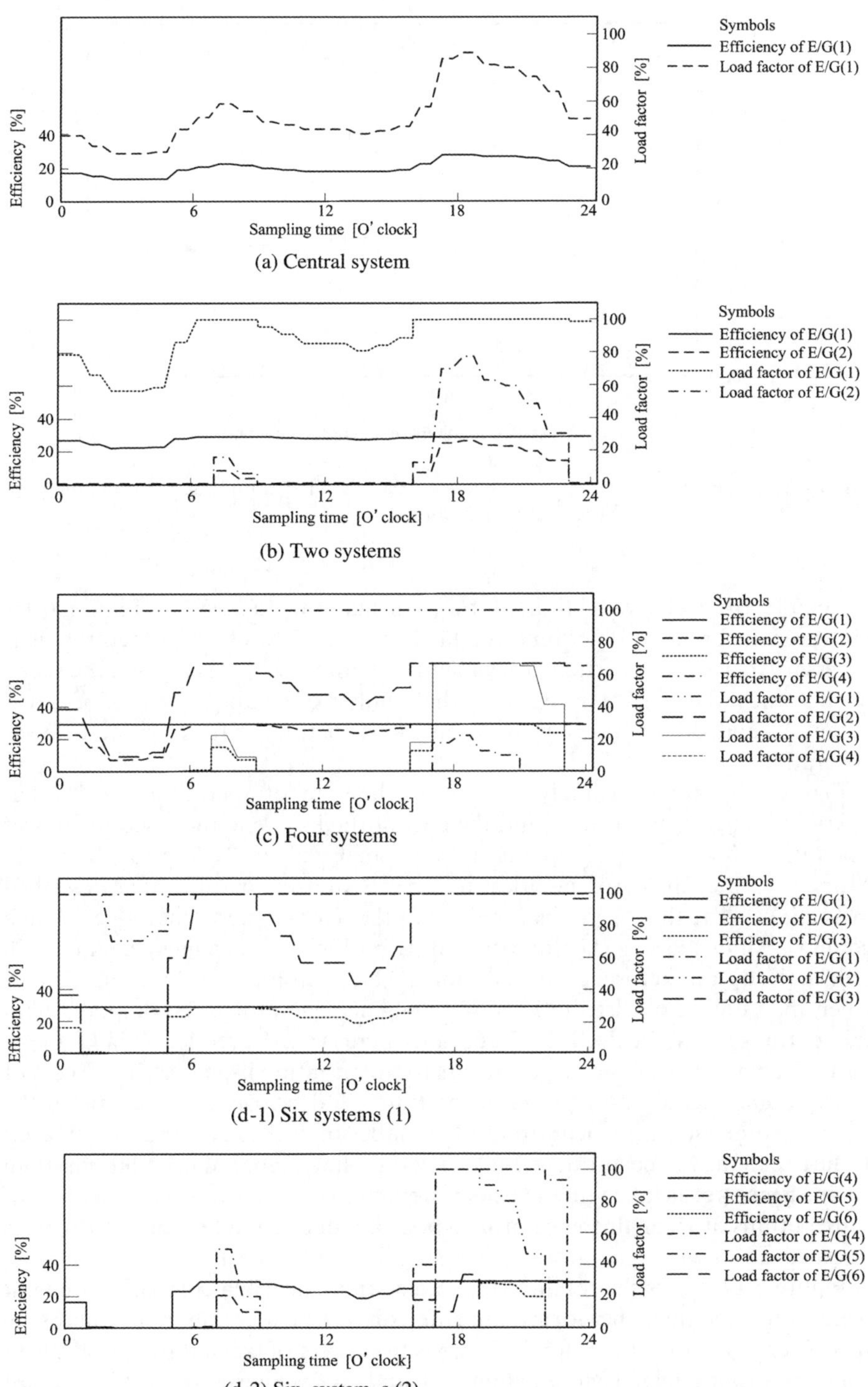

Figure 6.14 Analysis results of efficiency and load factor; (a) Central system; (b) Two systems; (c) Four systems; (d-1) Six systems (1); (d-2) Six systems (2).

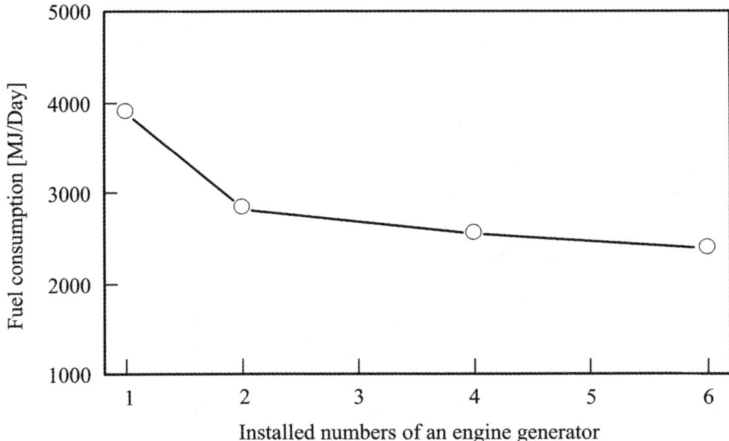

Figure 6.15 Result of the relation between the installed numbers of an engine generator, and the fuel consumption.

calculable from the reduction of fuel consumption shown in Figure 6.15. However, in order to determine the facility cost of an engine generator, it is necessary to assume a particular heat supply method, which is beyond the scope of this study. This chapter considers the number of engine generators for distribution, and the relationship between the reduction of fuel cost and power generation.

Figure 6.16 shows the analysis results of the years of operation of IMG that distributes engine generators, and the cost of the kerosene fuel used for power generation. The years of operation were analyzed as 3, 5, 8 or 10 years. Moreover, since the kerosene prices in the U.S. and Japan differed as described in Section 6.4.1, they were analyzed using the average kerosene price of each country. An increase in the number of distributions of an engine generator will reduce the cost of kerosene consumption as shown in Figure 6.15. In particular, when the number of distributions of an engine generator is 2, compared with the central system, the decline is large, and there is a 27% reduction in kerosene cost. If the number of engine generators for distribution is increased, the cost of the kerosene fuel used for power generation will decrease, but the ratio of the effect becomes so insignificant that the number of engine generators increases. If the rise in facility costs due to the increase in the number of installations of an engine generator is taken into consideration, even if it determines the number of distributions of an engine generator to be 2, a reduction in kerosene fuel cost is attained.

Figure 6.17 shows the result of the reduction value when comparing the cost of the kerosene fuel when operating IMG for 1 year with the cost of kerosene fuel when operating IMG for 3, 5, 8, and 10 years. For example, when IMG comprises four engine generators and it operates for 8 years in the U.S., a cost reduction of the kerosene fuel of 43 000 dollars is predicted. Therefore, the price

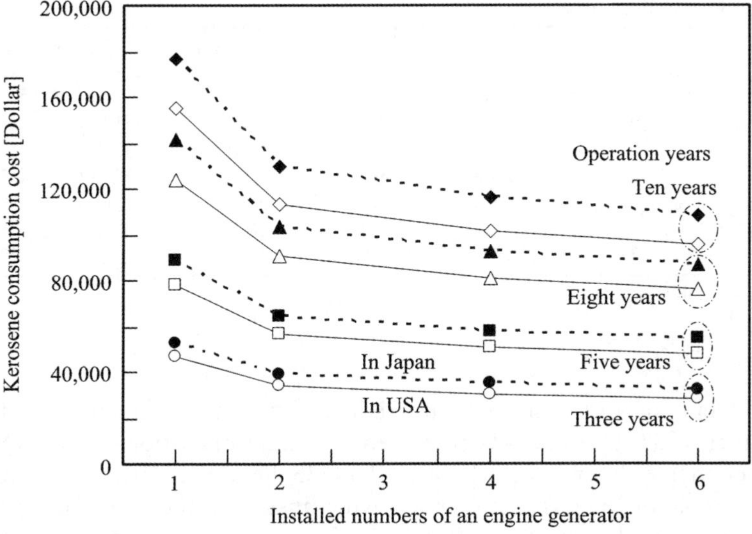

Figure 6.16 Analysis results of the number of installations of engine generators, and kerosene fuel cost.

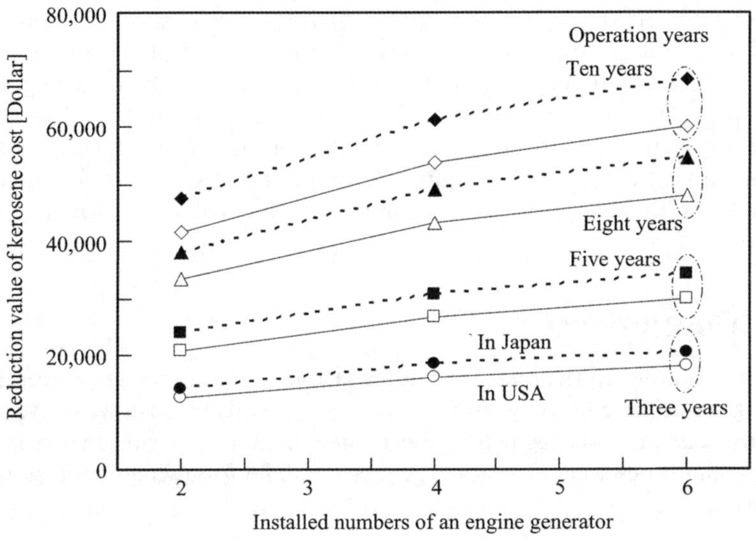

Figure 6.17 Analysis results of the number of installations of engine generators, and the reduction value of kerosene cost. Comparison with the operation cost for one year.

per engine generator is 10 750 dollars (43 000 dollars/four sets). However, considering the facility cost of IMG, the maintenance cost, the cost of equipment of the heat system, *etc.* will further reduce facility costs and maintenance costs per engine generator. Under the same conditions, if the number of

Table 6.3 Cost comparison.[a]

Installed number	*Generation capacity (kW)*	*Fuel cost*	*Initial cost*	*Maintenance cost*	*Sum of cost 1st year*	*Sum of cost 5th year*
1	20	17 700	33 900	340	51 940	187 205
2	10	12 900	11 900	510	25 310	72 410
3	6	12 300	9300	660	22 260	58 815
4	5	11 600	10 200	850	22 650	62 620
5	4	11 200	11 020	1020	23 240	66 325
6	3	10 900	11 900	1170	23 970	70 430

[a]costs in US Dollars

arrangements of an engine generator is two sets, a cost reduction of the kerosene fuel of 33 400 dollars is predicted assuming operation for 8 years. The price per engine generator is 16 700 dollars (33 400 dollars/two sets). Table 6.3 is an example of the cost comparison when installing the proposed system in Japan. However, equipment cost, installation cost, and a subsidy are included in the initial cost. The equipment cost included in the initial cost was obtained by referring to the data of business-use cogeneration and gas-engine cogeneration.[10,11] Maintenance cost is decided by the installed number and capacity of an engine generator. Initial cost and maintenance cost in Table 6.3 differ according to the product and the usage. If the number of installations of an engine generator changes, the power-generation capacity of an engine generator will also change. Therefore, although it is necessary to take the difference in power-generation capacity and the relationship of the facility cost into consideration, the number of engine generators for distribution is expected to be 3 or 4 sets; the analysis results of Figure 6.16 and Table 6.3 provide the same prediction.

6.6 Conclusions

The relation between the number of installations of an engine generator and the power-generation efficiency and the power generation cost was investigated by numerical analysis regarding the independent microgrid (IMG) that distributes small diesel-engine power generators. The following conclusions were obtained.

(1) If the engine generator linked to IMG is distributed and arranged, the power-generation efficiency of the whole system will improve compared with the central system. However, an engine generator with short hours worked appears. If the life of the engine generator is taken into consideration, it is necessary to replace the starting order of the engine generator within a definite period.

(2) The relationship between power generation and the exhaust heat of an engine generator is nonlinear. Therefore, operation fit for the power

and heat-demand pattern may be realizable by controlling the distribution of the load of the distributed engine generator.

(3) There is a reduction in the fuel for power generation, so that there are many engine generators distributed and arranged. However, if the facility cost and maintenance of an engine generator by distribution are taken into consideration, the number of installations of an engine generator would be 3 or 4 sets.

The fall in power-generation efficiency at the time of partial load operation of IMG was a problem. This chapter proposes a more efficient method of IMG with load fluctuation. Application of IMG using renewable energy is also considered.

Acknowledgments

This work was partially supported by a Grant-in-Aid for Scientific Research(C) from JSPS.KAKENHI (17510078).

Nomenclature

E/G: engine generator
IMG: independent microgrid
E_{eg}: power of an engine generator kW
R_{eg}: load factor %
η_{eg}: generation efficiency %

References

1. Z. Alibhai, R. Lum, A. Huster, W. A. Gruver, and D. B. Kotak, Coordination of Distributed Energy Resources, *IEEE International Conference on Systems, Man and Cybernetics, Vol. 2*, 2004, H. G. Stassen eds., Hague, Netherlands, 1990–1995.
2. A. Carlos and A. Hernandez, Fuel Consumption Minimization of a Microgrid, *IEEE Trans. Indus. Appl.*, 2005, **41**(3), 673–681.
3. S. Abu-Sharkh, R. J. Arnold, J. Kohler, R. Li, T. Markvart, J. N. Ross, K. Steemers, P. Wilson and R. Yao, Can Microgrids Make a Major Contribution to UK Energy Supply? *Renew. Sustain. Energy Rev.*, 2006, **10**(2), pp. 78–127.
4. H. Robert, Microgrid: A Conceptual Solution, *Proceedings of the 35th Annual IEEE Power Electronics Specialists Conference*, 2004, Aachen, **6**, 4285–4290.
5. M. Tanrioven, Reliability and cost-benefits of adding alternate power sources to an independent microgrid community, *J. Power Sources*, 2005, **150**, 136–149.

6. Yanmar Co., Ltd., Cogeneration product lineup, 2006, http://www.yanmar.co.jp/products/mgasc/images/lineup.jpg.
7. Fuji Futures Co., Commodity futures "market total information, 2005, http://www.fuji-ft.co.jp/selection/toyu/index.htm.
8. National Astronomical Observatory, Chronological Scientific Tables CD-ROM, *Rika Nenpyo* 2003, Maruzen Co., Ltd.
9. K. Narita, Research on Unused Energy of Cold Region Cities and Utilization for District Heat and Cooling, 1996, *Ph.D. thesis, Hokkaido University* (in Japanese).
10. Yanmar Co., Ltd., Cogeneration product lineup, 2006, http://www.yanmar.co.jp/products/mgasc/cogene03.htm.
11. Osaka Gas Co., Ltd., Osaka Gas home use gas appliance synthesis information site, 2006, http://www.g-life.jp/html/scene/cogeneration/ecowill/point/point09.html.

CHAPTER 7

PEFC/Green Energy Compound System (1) – Operation Planning of a PEFC and Photovoltaics with Prediction of Electricity Production Using GA and Numerical Weather Information

SHIN'YA OBARA

7.1 Introduction

An energy supply system using a microgrid constitutes the optimal system for energy demand. Therefore, its use as a clean energy-supply technique is expected to spread.[1–3] A microgrid using a PEFC (proton-exchange membrane fuel cell) may become the mainstream of future distributed energy. In addition, the application of green energy to a microgrid is desired. Accordingly, this chapter examines the PEFC and a photovoltaics compound system. Power can be supplied to a grid from each PEFC and photovoltaic component in this system. The hydrogen supply method to the PEFC assumes that the steam reforms the LPG (liquefied petroleum gas). However, the power generation output characteristics and PEFC exhaust heat with a steam reformer are

RSC Energy Series No. 3
Compound Energy Systems: Optimal Operation Methods
By Shin'ya Obara and Arif Hepbasli

Published by the Royal Society of Chemistry, www.rsc.org

nonlinear with a load factor.[4] Furthermore, although the power and exhaust heat of the proposed system are utilized effectively, battery installation and a heat-storage tank are planned. Consequently, the operation plan of the proposed microgrid must be optimized as a nonlinear system considering electricity and heat storage. Concerning operation optimization of a nonlinear system with heat storage, we have summarized the use of a GA (genetic algorithm).[5–7] In addition, it is necessary to predict unstable photovoltaic electricity production for every sample time while optimizing operation of a compound microgrid with a PEFC and photovoltaics. Accordingly, numerical weather information (NWI) is used to predict photovoltaic electricity production.[8,9] Anyone can obtain NWI in Japan through the Internet. However, there is an error in the photovoltaic electricity production calculated using NWI compared to using the actual meteorological data. Consequently, the operation plan of the system using the NWI differs from operation under actual weather conditions. The cause of this difference in operation is not addressed in this chapter. Instead, the relation between the NWI error and the operation results of the system is clarified. It is shown that installing the operation optimization algorithm using NWI is important for operation of a PEFC microgrid with photovoltaics. The objective of this study is to develop an analysis algorithm to optimize operation of a PEFC microgrid with green energy.

7.2 System Configurations

7.2.1 PEFC and Photovoltaics Compound Microgrid

Figure 7.1 shows a scheme of a compound microgrid with PEFC and photovoltaics. The compound microgrid consists of a power system and a heat system. Here, the power system is not connected with a commercial power system. The power from a PEFC and a solar cell can be supplied simultaneously to a microgrid. Moreover, these power sources can accumulate electricity using a battery. The hydrogen (reformed gas) supplied to a PEFC is produced from LPG (liquefied petroleum gas) using a steam reformer.

Each piece of equipment of the power system and the heat system is operated by a system controller. The photovoltaic electricity production for every sample time in a target day is predicted using the NWI (the amount of solar radiation and outdoor air temperature) obtained by the system controller at 0:00 on the target day. Based on this prediction result, the optimal system operation on the target day is planned by the system controller. The objective given to the system controller is to minimize fuel (LPG) consumption. As Figure 7.1 shows, fuel is consumed by a PEFC and a boiler in the proposed system. The optimization analysis of the operation plan in this chapter considers operation of a power system and a heat system. The NWI used for analysis is the information obtained at 0:00 on a target day. Therefore, the NWI does not match actual meteorological data. If a system is employed according to the first optimization plan (the plan is made at 0:00 on the target day), then depending on the magnitude of

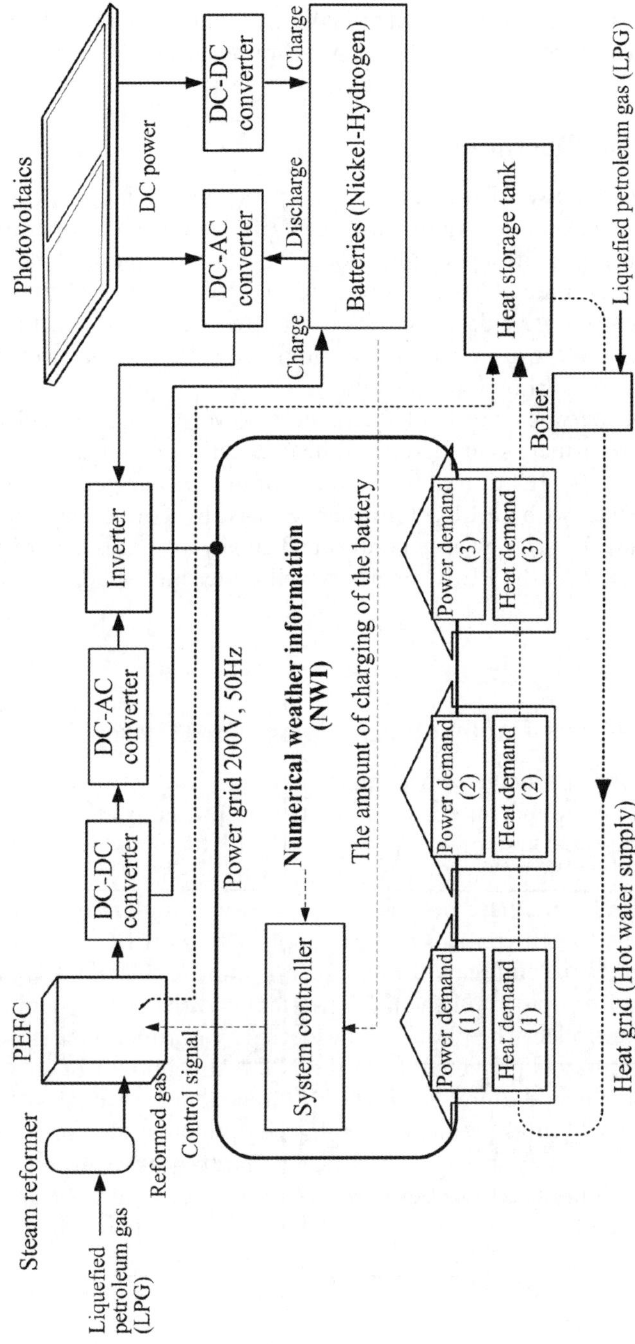

Figure 7.1 System scheme.

this error, the fuel consumption may get worse. For example, the operation hours of a PEFC and a boiler may be extended under the actual weather conditions. Investigating the relation between the NWI error and system fuel consumption evaluates the operation optimization algorithm using NWI.

7.2.2 System Operation

Figure 7.2 shows power demand (a) on a representative day, heat demand (d), electricity production and exhaust heat of PEFC (b), (e), the operation model of a battery (c) and the operation model of heat storage and the boiler (f). Predicted photovoltaic electricity production based on the NWI obtained at 23:00 on a representative day is shown in Figures 7.2(a) and (d). Furthermore, the photovoltaic electricity production obtained under actual weather conditions at each time is shown in this figure.

The relation between the load factor and power-generation efficiency of a PEFC with a reformer, load factor and heat-generation efficiency is nonlinear. Figure 7.3 shows the relation of the load factor and power-generation efficiency of a home fuel cell with an LPG reformer released by Tokyo Gas Co. Ltd.[10] As shown in Figure 7.3(a), power-generation efficiency also falls according to load decreases. Therefore, the PEFC operates well under high load. Accordingly, as

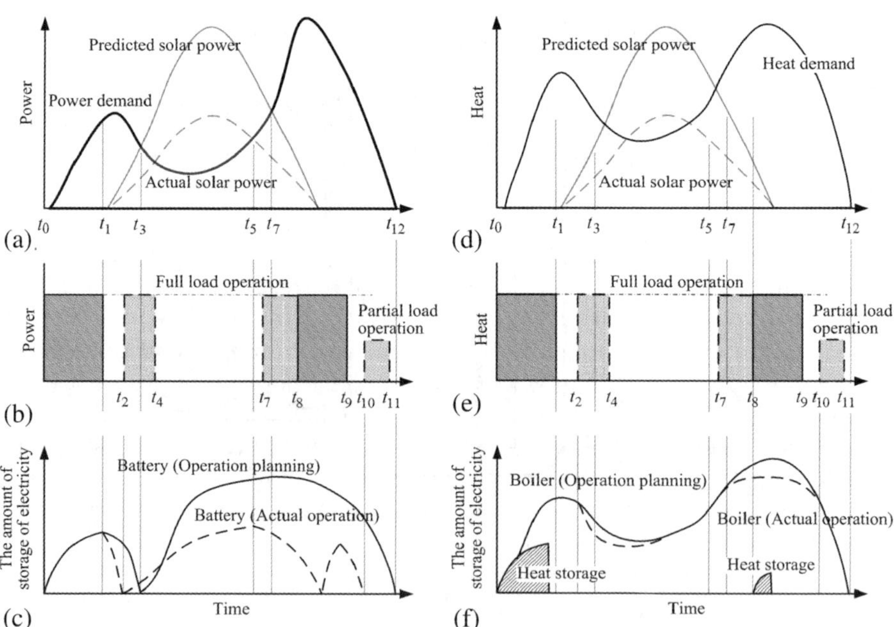

Figure 7.2 System operation planning of the compound microgrid installation of a PEFC and photovoltaics with prediction of the production of electricity.

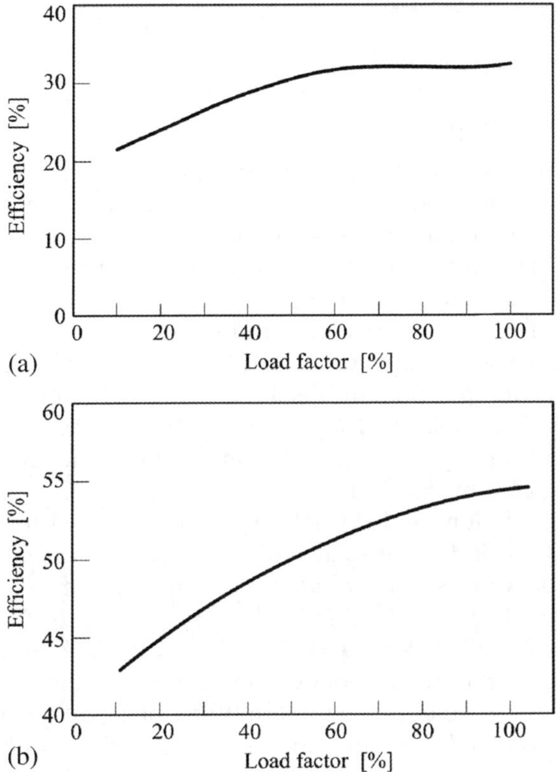

Figure 7.3 Efficiency characteristics of the PEFC with steam reformer; (a) Power-generation efficiency; (b) Heat-generation efficiency.

shown in Figure 7.2(b), it is desirable to operate the PEFC near its maximum efficiency point. When the PEFC is operated near the maximum efficiency point, exhaust heat of the PEFC is an output characteristic, shown in Figure 7.2(e). Operation of the PEFC in the periods from t_0 to t_1 and t_8 to t_9 in Figures 7.2(b) and (e) are the optimal operation plan based on predicted solar power shown in Figures 7.2(a) and (d). On the other hand, operation of PEFC in the periods from t_2 to t_4, t_7 to t_8 and t_{10} to t_{11} in Figures 7.2(b) and (e) covers the photovoltaic shortage compared to the actual solar power (the actual solar power is smaller than the predicted solar power), as shown in Figures 7.2(a) and (d). When there is little actual solar power compared to the predicted solar power, additional PEFC operation is required. In this case, as shown in Figures 7.2(c) and (f), battery, heat-storage tank and boiler operations change. Accordingly, the relation between the magnitude of the difference of the predicted solar power and the actual solar power, and the fuel consumption of the system is investigated. By considering this result, the influence of the NWI error on the system operation plan can be identified.

7.3 Analysis Method

7.3.1 Power System

7.3.1.1 Photovoltaics

In this chapter, installation of polycrystalline silicon solar modules of area S_s is assumed. The average production of electricity $P_{s,t}$ of the solar module from sample time t to $t+1$ on a representative day is calculated as shown in Figure 7.1. R_T in eqn (7.1) is the temperature coefficient, and when the temperature $T_{c,t}$ of the solar cell rises, power-generation efficiency will fall. T_o is a reference temperature, and η_s is the power-generation efficiency under T_o. The temperature $T_{c,t}$ of the solar cell is calculated from the specific heat of the polycrystalline silicon and the amount of solar radiation at sampling time t. When the intensities of direct solar and sky solar radiation are expressed by $H_{D,t}$ and $H_{M,t}$, respectively, among the solar radiation input into the acceptance surface, $P_{s,t}$ will be calculated by eqn (7.1).

Direct solar insolations and sky solar radiation are used for power generation in a flat solar cell. Global-solar-radiation intensity, direct solar radiation intensity, and horizontal sky solar radiation intensity at time t ($t=0.1, 2, \ldots,$ 23) are expressed with $I_{H,t}$, $I_{D,t}$ and $I_{M,t}$, respectively. $I_{H,t}$ and $I_{D,t}$ can be determined from the NWI. Moreover, $I_{M,t}$ can be calculated using $I_{H,t}$ and $I_{D,t}$. The incidence angle θ to the acceptance surface of sunlight is calculated using eqn (7.2). Here, φ, δ, and ω show the latitude of a setting point, the solar celestial declination, and hour angle, respectively, while eqn (7.3) is a calculation formula for the sky solar radiation component $H_{D,t}$.

$$P_{s,t} = S_s \cdot \eta_s \cdot (H_{D,t} + H_{M,t}) \cdot \{1 - (T_{c,t} - T_o) \cdot (R_T/100)\} \tag{7.1}$$

$$\sin\theta = \cos\varphi \cdot \sin\delta - \sin\varphi \cdot \cos\omega \cdot \cos\delta \tag{7.2}$$

$$H_{D,t} = I_{D,t} \cdot \cos\theta \tag{7.3}$$

Equation (7.4) calculates the incidence sky solar radiation component $H_{M,t}$ of the solar cell. The first term on the right-hand side of eqn (7.4) is the air solar radiation component; the second term is the reflective solar radiation component; β is the angle of gradient of the acceptance surface by eqn (7.5); and ρ is the reflection factor of the ground.

$$H_{M,t} = I_{M,t} \cdot \frac{1+\cos\beta}{2} + \rho \cdot I_{H,t} \cdot \frac{1-\cos\beta}{2} \tag{7.4}$$

$$\cot\beta = \cos\varphi \cdot \cot\omega + \sin\varphi \cdot \operatorname{cosec}\omega \cdot \tan\delta \tag{7.5}$$

7.3.1.2 Power Balance

Equation (7.6) is a power-balance equation. $P_{fc,t}$, $P_{pv,t}$ and $P_{bt,t}$ on the left-hand side in the equation are the PEFC power, photovoltaic power, and battery power, respectively. $P_{need,t}$, $P_{btc,t}$, $P_{loss,t}$ on the right-hand side in the equation represent power demand, the amount of battery charge, and loss of power, respectively. Charge-and-discharge loss of a battery is included in the power loss $P_{loss,t}$.

$$P_{fc,t} + P_{pv,t} + P_{bt,t} = P_{need,t} + P_{btc,t} + P_{loss,t} \tag{7.6}$$

7.3.2 Heat Balance

Equation (7.7) is a heat-balance equation.

$H_{fc,t}$, $H_{bl,t}$ and $H_{st,t}$ on the left-hand-side in the equation are the heat power of a fuel cell, a boiler, and a heat-storage tank, respectively. $H_{need,t}$, $H_{sts,t}$ and $H_{loss,t}$ on the right-hand side of the equation are heat demand, the amount of heat storage, and the heat loss, respectively. Heat storage loss is included in the heat loss $H_{loss,t}$ on the right-hand-side of the equation.

$$H_{fc,t} + H_{bl,t} + H_{st,t} = H_{need,t} + H_{sts,t} + H_{loss,t} \tag{7.7}$$

7.3.3 Optimal Analysis Using GA

7.3.3.1 Objective Function

If $P_{fc,t}$ in eqn (7.6) and $H_{bl,t}$ in eqn (7.7) are determined, the heating value of LPG $Q_{fuel,t}$ consumed by a compound microgrid is calculable. Here, the amount of fuel with the output power of $P_{fc,t}$ and $H_{bl,t}$ is decided by the PEFC power-generation efficiency and the thermal efficiency of the boiler. Equation (7.8) defines the objective function in this study. The objective function minimizes the system fuel consumption $Q_{system,day}$ on one day. The fuel consumption $Q_{fuel,t}$ of the system from sample time t to $t+1$ is the sum of the fuel consumption $Q_{fc,t}$ of a fuel cell, and the fuel consumption $Q_{bl,t}$ of a boiler.

$$Q_{system,day} = \sum_{t=0}^{23} Q_{fuel,t} = \sum_{t=0}^{23} (Q_{fc,t} + Q_{bl,t}) \tag{7.8}$$

7.3.3.2 Optimal Operation Planning Algorithm

In this study, the optimal operation plan of the proposal compound microgrid is analyzed using a GA. Figure 7.4 shows the operation optimization algorithm developed in this chapter, and the analysis flow is explained below.

(1) The energy-demand pattern data, equipment specifications, GA parameters, numerical weather data, efficiencies, initial conditions and system loss are used as input into a computer (system controller) in calculation (A) in Figure 7.4.

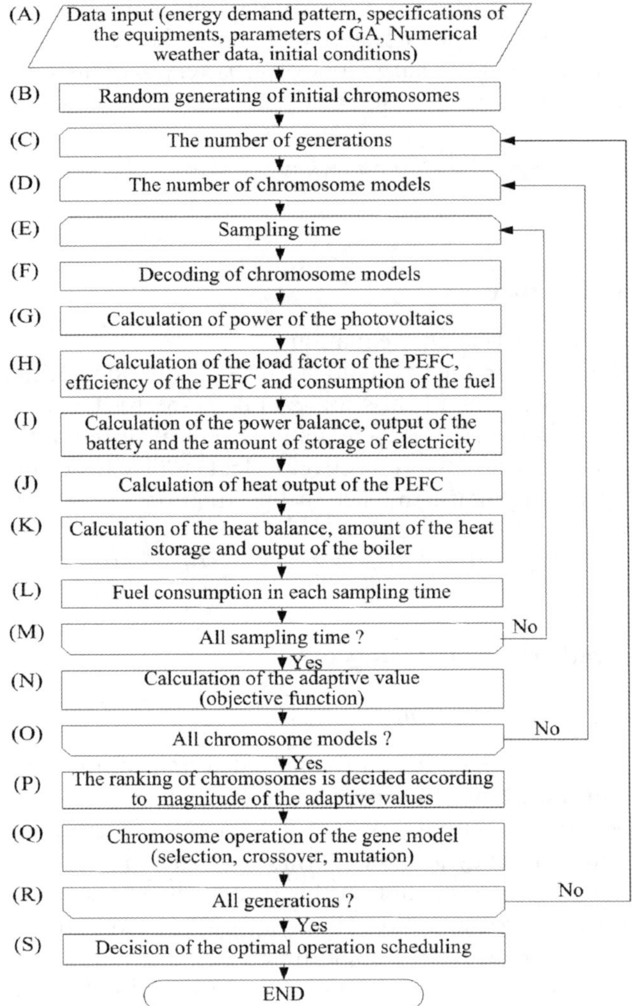

Figure 7.4 Optimal operation algorithm.

(2) In calculation (B), many initial generation chromosome models are generated at random. One individual of the chromosome model expresses PEFC operation and power. The PEFC operation is represented with a 1-bit binary number and the PEFC power is represented by a 14-bit binary number.
(3) In calculation (F), the PEFC power is determined by decoding the chromosome model. Furthermore, in calculation (G), the production of electricity of photovoltaics is calculated using NWI (Section 7.3.1.1).
(4) In calculations (H) through (K), battery, heat-storage tank, and boiler operations are planned based on the power balance and heat-balance equations (Sections 7.3.1.2 and 7.3.2).

(5) The fuel consumption is calculable from the amount of PEFC and boiler power. In calculation (L), these values are totalled, and the fuel consumption of the system in the sampling time *t* is determined.
(6) Calculations (E) through (M) are repeated from sampling time 0 to 23 for one chromosome model. In calculation (N), the adaptive value (namely, the objective function shown in eqn (7.8) of the chromosome model) is obtained from this result.
(7) The adaptive value of all the chromosome models is decided by repeating calculations (D) through (O). The ranking of the chromosome models is decided according to the magnitude of the adaptive value of each chromosome (calculation (P)).
(8) The chromosome models with low adaptive value are selected, and they are exchanged for the new randomly generated model. Moreover, the genetic manipulation of crossover and mutation is added based on the probability given in calculation (A) concerning the chromosome models with high adaptive value (calculation (Q)).
(9) Calculations (D) through (Q) (repeated calculation of calculations (C) to (R)) are repeated for a defined number of generations. In the last generation's chromosome group, the solution with the highest adaptive value is chosen to be the optimal system operation plan (calculation (S)).

7.4 Case Analysis

7.4.1 Equipment Specifications

The equipment specifications for the case analysis of the PEFC and photovoltaics compound microgrid are shown in Table 7.1. The microgrid assumes that the equipment is installed in Sapporo, Japan (latitude 43.062-degree north and longitude 141.354-degree east, a cold and snowy area).

Table 7.1 Specification of equipment.

PEFC maximum power	3 kW
PEFC performance	Fig. 3
Solar cell type	Multicrystalline silicon
Area of solar cell	60.0 m^2
Maximum generation efficiency of the solar cell	16.4%
Temperature coefficient of the solar cell	0.4%/K
Battery type	Nickel-hydrogen
Amount of battery self-discharge	10%/h
Efficiency of converter	95%
Efficiency of inverter	95%
Loss of heat storage	5%/h
Efficiency of boiler	90%

(1) PEFC with a reformer.

The maximum PEFC power with a reformer is 3 kW, and this performance is shown Figure 7.3.

(2) Photovoltaics.

The maximum efficiency and photovoltaic temperature coefficients are 16.4% and 0.4%/K, respectively. These values are general facility values used in Japan. The solar panel is installed in the roof, with a slope of 30-degrees facing south. Moreover, the solar cell area is set to 60.0 m^2.

The area of the general solar cell installed into individual houses in Japan is usually 25 m^2 to 40 m^2 (for a solar cell with a 3-kW to 5-kW capacity).

(3) Battery, converter and inverter.

The self-discharge of a battery is set to 10% per hour. The converter and inverter efficiencies are both set to 95%.

(4) Heat-storage tank and boiler.

Heat dissipation loss of a heat-storage tank is set to 5% per hour, and the boiler efficiency is set to 90%.

7.4.2 GA Parameters

The GA parameters in the proposed algorithm are shown in Table 7.2. These values were chosen by repeating trial and error so that the convergence solution was as stable as possible. Since the convergence solution (analysis result) has dispersion for every analysis, the optimal solution is obtained by repeating the same analysis.

7.4.3 Energy Demand Pattern

Power and heat are supplied to three individual houses in Sapporo, Japan using the proposed microgrid. Figure 7.5 shows the power and heat demand on a representative day every month.[11] There is no cooling load and heating is included in the heat load. Therefore, the power load pattern on a representative day of every month does not vary significantly throughout the year. On the other hand, the magnitude of demanded heat varies greatly between the summer season and winter season.

7.4.4 Error of the NWI

Various error characteristics of the NWI can be considered. However, in the investigated case, various NWI errors were not found. Accordingly, the error

Table 7.2 GA parameters.

The number of chromosome models	10000
Generation number	100
Probability of mutation	10%
Selection	50% of a low rank is replaced

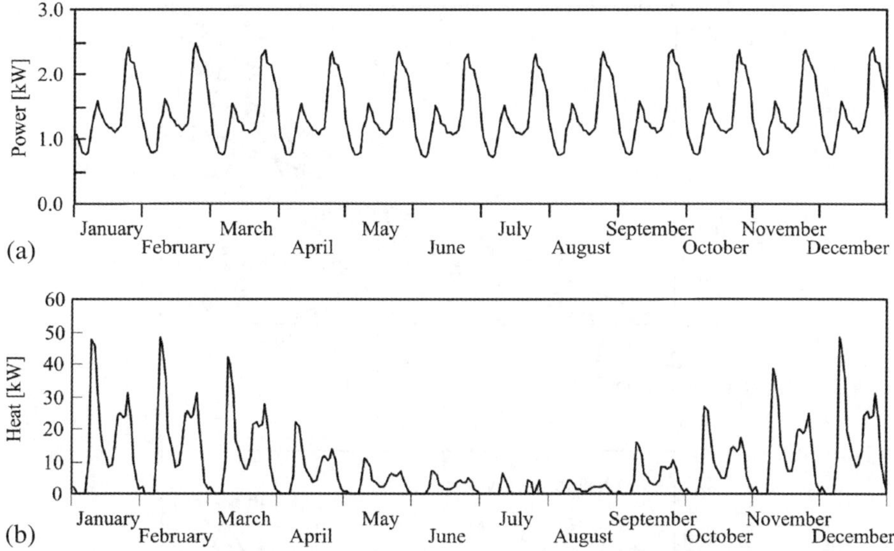

Figure 7.5 Energy-demand model. Load patterns of three individual houses on representative days every month in Sapporo; (a) Power-demand model; (b) Heat-demand model.

pattern of the following two types is installed in this study. Figure 7.6 shows the error pattern of the two types used to analyze the proposed algorithm (Figure 7.4). As shown in Figure 7.6(a), a linear error and a quadratic error are installed as error patterns. Target day operation plans are determined by the NWI at 0:00 on the target day. In this chapter, the error that is proportional ($error_t = const1 \cdot t$) to time is defined as the linear error. On the other hand, the error that follows ($error_t = const2 \cdot t^2$) a secondary curve relative to time is defined as the quadratic error. Here, the integrated values of the two error types are set equal to each other. Therefore, *const*1 and *const*2 were fixed so that Area A and Area B (shown in Figure 7.6(a)) might become equal (in Figure 7.6(a), Area A = Area B = 1.0). The common characteristic of these error patterns is the increase of the NWI error as time increases. Moreover, as shown in Figures 7.6(b) and (c), fluctuation errors of $\pm 20\%$ and $\pm 40\%$ at random are added to the two error types. These fluctuation errors simulate the instability of the solar insolation data.

7.5 Results and Discussion

7.5.1 Operation Planning

Figure 7.7 shows the results of the system operation plan optimization analysis on representative February days (winter). Figures 7.7(a) and (b) are the optimal operation plans for a power system and a heat system, respectively. Moreover,

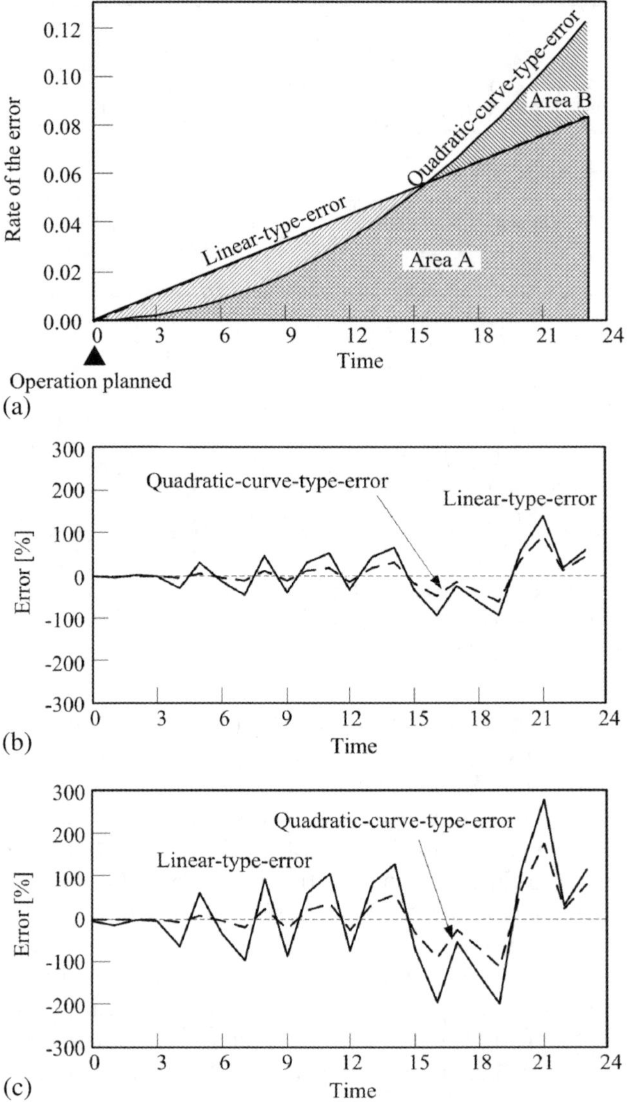

Figure 7.6 Error function and error function with random error of the numerical weather information; (a) Error function; (b) Error of the numerical weather information ±20%; (c) Error of the numerical weather information ±40%.

Figure 7.7(c) shows the fuel consumption plan in this case. The fuel consumption is the sum total of each value of a PEFC with a reformer and a boiler. Figure 7.8 shows the operation results of the system at the time of the linear error and quadratic error on the NWI. Similarly, Figure 7.9 shows the results of the system operating plan optimization analysis on representative August days

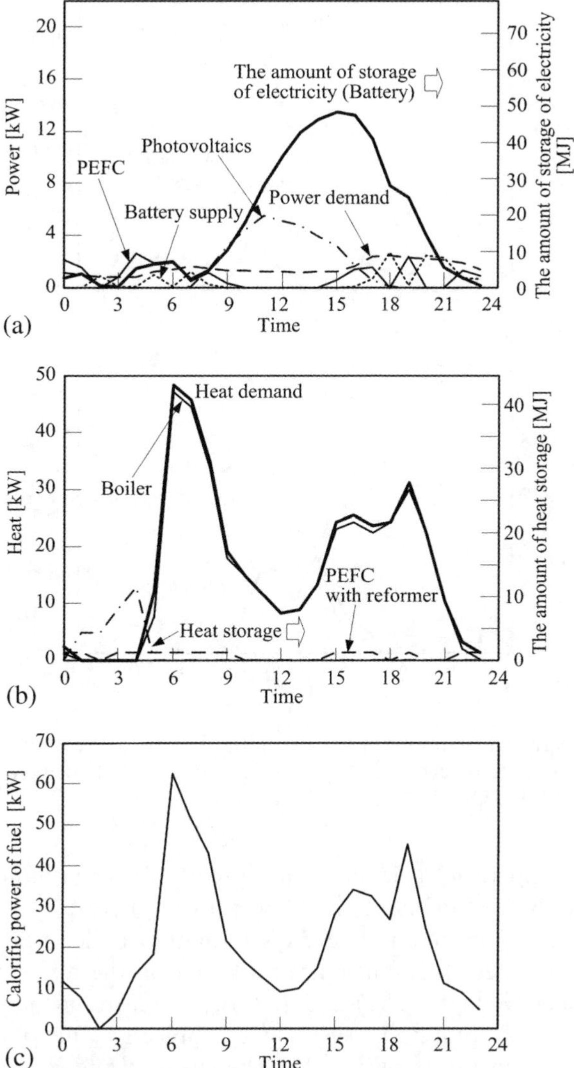

Figure 7.7 Analysis results of the operation planning of the proposal microgrid in February; (a) Operation planning of the power system; (b) Operation planning of the heat system; (c) Fuel consumption.

(summer season). Figure 7.10 shows the operation result with the two error types on the NWI.

The battery operation plans shown in Figure 7.7(a) and Figure 7.9(a) differ greatly for each month. Accordingly, the amount of maximum electricity storage in August is clearly large compared with that in February. This is because of the difference in the photovoltaic power generation in February and August. Moreover, when Figure 7.7(b) is compared with Figure 7.9(b), the ratios of the

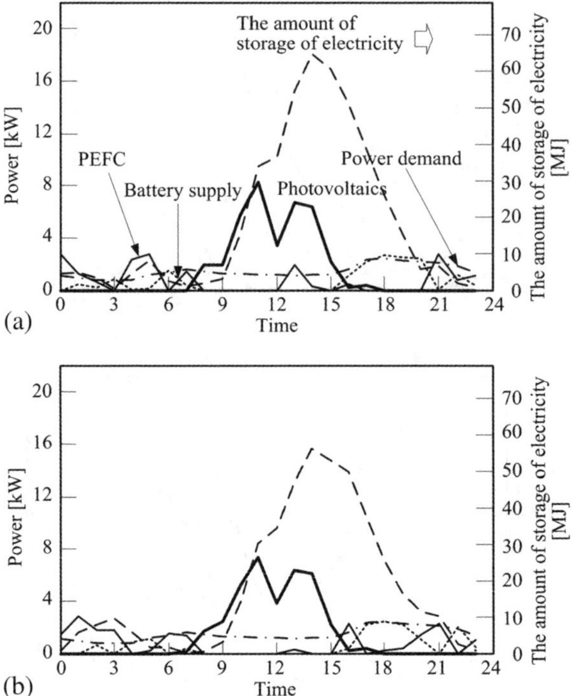

Figure 7.8 Operation planning in the case of a numerical weather information with two types of error ±40%. Power system, in February. (a) Linear-type error; (b) Quadratic-curve-type error.

PEFC exhaust heat to heat demand vary greatly for each month. The PEFC exhaust heat to the heat demand ratio is very low in February. As a result, heat supply on February representative days is mainly boiler heat. The summer season has little system fuel consumption based on the difference in the heat power of a boiler (Figures 7.7(c) and 7.9(c)). Therefore, if the proposed compound microgrid is optimized based on the objective function in eqn (7.8), power should be generally optimized in the summer and heat should be generally optimized in the winter.

7.5.2 Influence of the Numerical Weather Information Error

The relation between the NWI error and the power system operation results is investigated (Figures 7.8 and 7.10). If an error is included in the NWI, the amount of storage of electricity will increase sharply for any month. From this result, the time shift of power is conjectured to perform an important role for optimizing system operation with NWI error. Accordingly, because the operation plan is strongly influenced by battery capacity setup, it is thought that the fuel consumption of the system changes greatly.

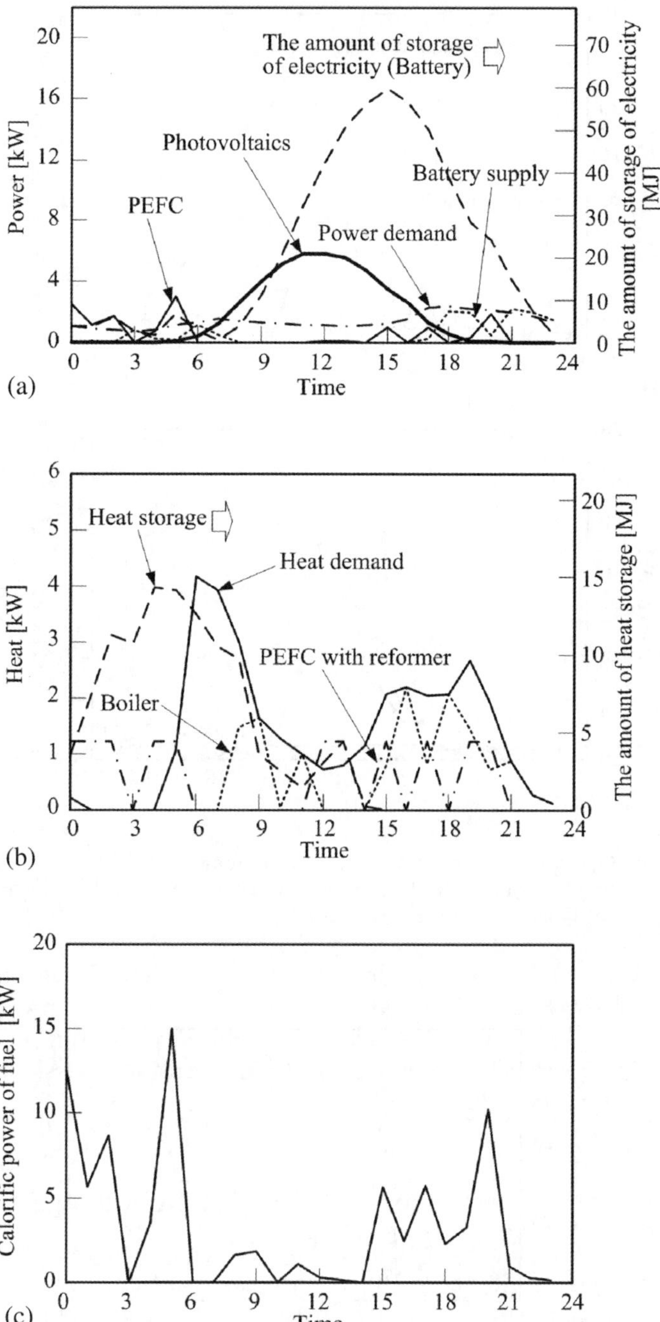

Figure 7.9 Analysis results of the operation planning of the proposal microgrid in August. (a) Operation planning of the power system; (b) Operation planning of the heat system; (c) Fuel consumption.

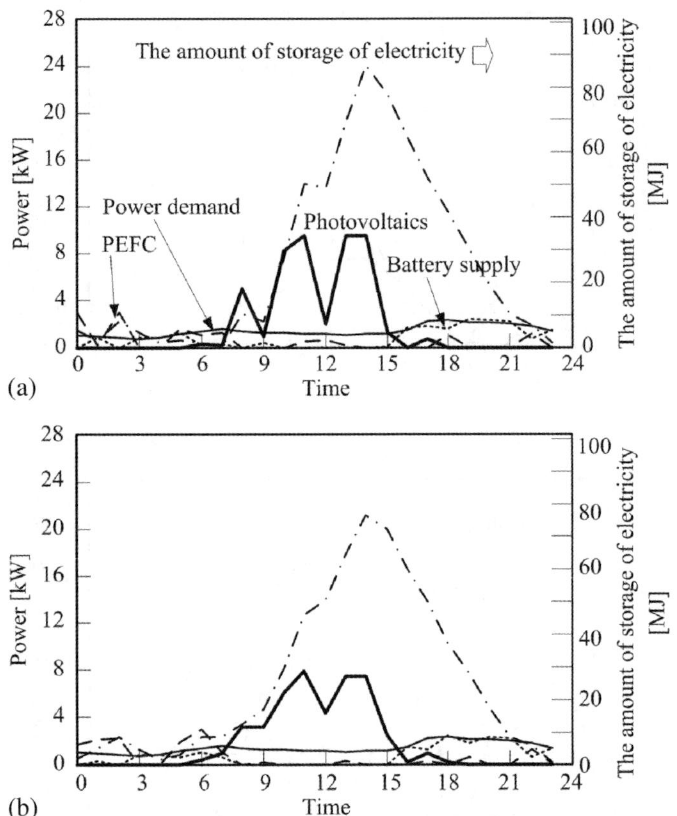

Figure 7.10 Operation planning in the case of a numerical weather information with two types error $\pm 40\%$. Power system, in August. (a) Linear-type error; (b) Quadratic-curve-type error.

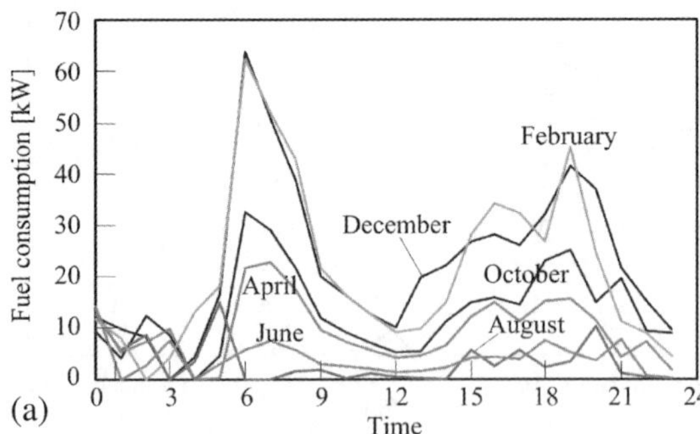

Figure 7.11 Analysis results of the fuel consumption; (a) Optimal operation planning; (b) In the case of a numerical weather information with 20% error; (c) In the case of a numerical weather information with 40% error.

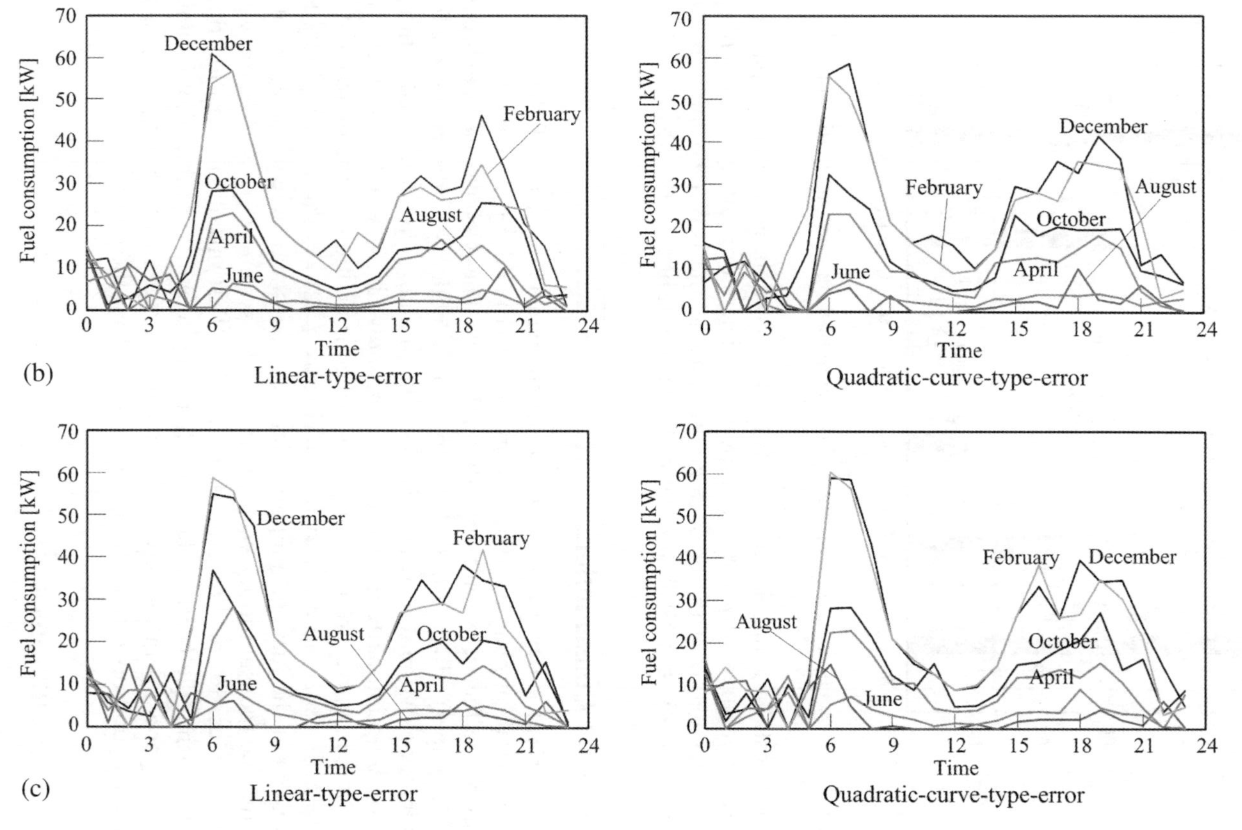

Figure 7.11 Continued.

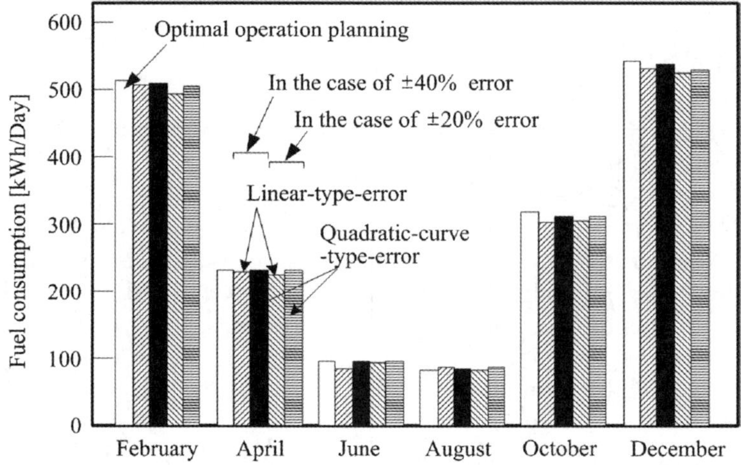

Figure 7.12 Analysis results of the fuel consumption on a representative day.

When installing the linear error into the NWI, large battery capacity is required compared with the quadratic error. Therefore, the system operation method changes with the error characteristics of the NWI. To minimize the battery capacity, the NWI quadratic error is desirable.

7.5.3 Fuel Consumption

Figure 7.11(a) shows the fuel consumption plan in the case of the optimal system operation plan. Moreover, Figures 7.11(b) and (c) show the operation results of the fuel consumption with NWI error. The winter season (February and December) with the large heat power of a boiler requires significant fuel consumption. Moreover, when Figures 7.11(b) and (c) are compared with the fuel consumption pattern for every month, shown in Figure 7.11(a), there is a clear difference. Accordingly, the total fuel consumption on the representative day was calculated about every month (Figure 7.12). As shown in Figure 7.12, the results of the fuel consumption plan in the optimal system operation plan and the fuel consumption when operating the system with NWI error were small in value. From this result, it is surmised that installing operation optimization using the NWI in the fuel cell microgrid with unstable photovoltaics achieves good operation. Even if it includes error in the NWI, the system maintains good operation and hardly suffers from the error.

7.6 Conclusions

In this chapter, the photovoltaic electricity production was predicted using numerical weather information (NWI), and a system operation optimization algorithm based on NWI was proposed. The proposed algorithm uses a GA (genetic algorithm), and optimizes the system operation plan. However, since error exists between NWI and meteorological data in real time, the operation of an actual system differs from the optimal operation plan defined beforehand. Accordingly, in this chapter, the relation between the error characteristic of the NWI and fuel consumption of the system were clarified. Moreover, the following was concluded. First, when the proposal compound microgrid is installed in a cold region and optimized, power is mainly optimized in the summer and heat is primarily optimized in the winter. Secondly, for system operation with NWI error, the power time shift has an important role. Accordingly, the operation plan changes greatly with the magnitude of the battery capacity. As a result, system fuel consumption varies greatly from month to month. Lastly, high-performance operation can be achieved by installing the operation optimization method based on the NWI into the fuel cell microgrid with unstable photovoltaics. Even if the error shown in this chapter is included in the NWI, the influence on the system fuel consumption is small.

Acknowledgements

This work was partially supported by a Grant-in-Aid for Scientific Research (C) from JSPS.KAKENHI (20560204).

Nomenclature

H: heat W
I_D: direct solar radiation intensity W/m^2
I_H: global-solar-radiation intensity W/m^2
I_M: horizontal sky solar radiation intensity W/m^2
H_D: intensity of direct solar W/m^2
H_M: intensity of sky solar radiation W/m^2
P: power W
Q: fuel consumption W
R_T: temperature coefficient %/K
S_s: area of the solar cell m^2
T: temperature K
T_c: temperature of the solar cell K
T_o: reference temperature K
t: sample time

Greek Symbols

β: angle of the acceptance surface gradient
δ: the solar celestial declination
φ: the latitude of a setting point
η_s: photovoltaics efficiency at T_o%
θ: incident angle to the sunlight acceptance surface
ρ: ground reflection factor
ω: hour angle

Subscripts

bl: boiler
bt: battery
btc: battery discharge
fc: PEFC with reformer
loss: energy loss
need: energy demand
pv: photovoltaics
s: solar module
st: heat storage
sts: heat storage output

References

1. Shin'ya Obara, Equipment plan of compound interconnection microgrid composed from diesel power plants and solid polymer membrane-type fuel cell, *Int. J. Hydrogen Energy*, 2008, **33**(1), 179–188.
2. Hirohisa Aki, Shigeo Yamamoto, Junji Kondoh, Tetsuhiko Maeda, Hiroshi Yamaguchi, Akinobu Murata and Itaru Ishii, Fuel cells and energy networks of electricity, heat, and hydrogen in residential areas, *Int. J. Hydrogen Energy*, 2006, **31**(8), 967–980.
3. Huang Jiayi, Jiang Chuanwen and Xu Rong, A review on distributed energy resources and MicroGrid, *Renew. Sustain. Energy Rev.*, 2008, **12**(9), 2472–2483.
4. S. Obara, Load response characteristics of a fuel cell microgrid with control of number of units, *Int. J. Hydrogen Energy*, 2006, **31**(13), 1819–1830.
5. Shin'ya Obara, Operating Schedule of a Combined Energy Network System with Fuel Cell, *Int. J. Energy Res.*, 2006, **30**(13), 1055–1073.
6. Shin'ya Obara, Equipment Arrangement Planning of a Fuel Cell Energy Network Optimized for Cost Minimization, *Renew. Energy*, 2007, **32**(3), 382–406.
7. Shin'ya Obara, The Exhaust Heat Use Plan when Connecting Solar Modules to a Fuel Cell Energy Network, *Trans. ASME, J. Energy Res. Technol.*, 2007, **129**(1), 18–28.
8. Online data service, GPV/GSM (Grid Point Value/GSM (Global Spectral Model)), http://www.jmbsc.or.jp/hp/online/f-online0a.html, Japan meteorological business support center, 2009.
9. Data of Japan Meteorological Agency, http://database.rish.kyoto-u.ac.jp/arch/jmadata/gpv-original.html, Kyoto University, 2009.
10. Development of home fuel cell cogeneration system in Tokyo Gas, ALIA News Vol. 89, http://www.alianet.org/homedock/15kinen/5-2.html, 2009.
11. K. Narita, *Research on unused energy of cold region cities and utilization for district heat and cooling*, Ph.D. thesis, Dept. Socio-Environmental Eng. Faculty of Eng., Hokkaido Univ. Sapporo, Japan, 1996.

CHAPTER 8

PEFC/Green Energy Compound System (2) – Overall Efficiency of a PEFC with a Bioethanol Solar Reforming System for Individual Houses

SHIN'YA OBARA

8.1 Introduction

Development of a distributed power supply with limited greenhouse-gas emission is a global issue of current interest and importance. PEFCs (polymer electrolyte fuel cells) are one candidate for a clean, distributed power supply. However, the environmental impact of fuel cells changes greatly depending on the method of hydrogen production. For example, a large quantity of CO_2 is discharged when using reforming methods that employ fossil fuels. Alternatively, a fuel cell system that uses the heat of a small solar collector for the steam reforming of bioethanol, a bioethanol-reforming system for fuel cells (FBSR system), has been examined.[1–3] Furthermore, there are researches on hydrogen production technology using solar energy currently (for example, refs 4–7). In this research, we have investigated the characteristics of the reformed gas production,[1] the production-of-electricity characteristic of the FBSR using the weather forecast,[2,3] and the relationship between solar insolation with fluctuation and a production-of-electricity characteristic.[8] This chapter continues these reports. The main objectives of this chapter are to compare the

RSC Energy Series No. 3
Compound Energy Systems: Optimal Operation Methods
By Shin'ya Obara and Arif Hepbasli

Published by the Royal Society of Chemistry, www.rsc.org

efficiency of the overall efficiency of the FBSR, and competition technology. If the overall efficiency of the proposal system is higher than competition technology, the trial production of the proposal system is effective. Moreover, in this study, fluctuation of solar radiation was taken into consideration to the reaction rate of steam reforming. There is no previous example of investigation of the solar reforming with solar radiation fluctuation. The conversion rate of the reforming reaction of ethanol/water vapor is influenced by the temperature of a catalyst layer and the space velocity of a process gas.[8] A pellet-type reforming catalyst is installed in the reactor of the FBSR. Reformed gas with a high hydrogen composition is output by supplying ethanol/water vapor to the reactor under the control of reactor temperature. However, because the heat source of a reactor is sunlight, the conversion rate of the fuel is affected by weather. Unsteady heat-transfer analysis was introduced into the catalyst layer in a previous examination.[8] From this analysis, the transient response characteristic concerning the temperature distribution in the catalyst layer and composition of the process gas were investigated. As a result, under weather conditions with high levels of solar radiation fluctuation over a short time, it became clear that a reforming reaction was not sufficient under the effect of a response delay. Moreover, it clarified the generation efficiency of the FBSR when taking into consideration the transient response characteristic of the reforming reaction. This chapter investigates the thermal output characteristic of the FBSR. The overall efficiency of the system is clarified in consideration of these results and the result of the last report. It is the objective of this chapter to highlight the differences between the FBSR and competing technologies, such as photovoltaic cells.

8.2 Material and Method

8.2.1 System Block Diagram

Figure 8.1(a) is a block diagram of the fuel cell system with a bioethanol solar reforming system (FBSR). The solar tracking system is introduced into two solar collectors. Moreover, each collector's collecting area is nearly 1.0 m^2. Bioethanol for the boiler and the bioethanol solution for reforming are contained in a fuel tank. The S/C (molar ratio of steam to ethanol[9]) of the bioethanol solution for reforming is 3.0. Two parabolic mirrors (solar collector) with a solar tracking system are introduced into the FBSR. The heat for fuel evaporation is condensed in solar collector A, while the heat for reforming of the fuel vapor is condensed in solar collector B.

8.2.2 Fuel and Reformed Gas System

(1) Flow of fuel and reformed gas.

The fuel for reforming (bioethanol solution) is supplied to the vaporizer installed in solar collector A from the fuel tank using a pump. The fuel vapor is

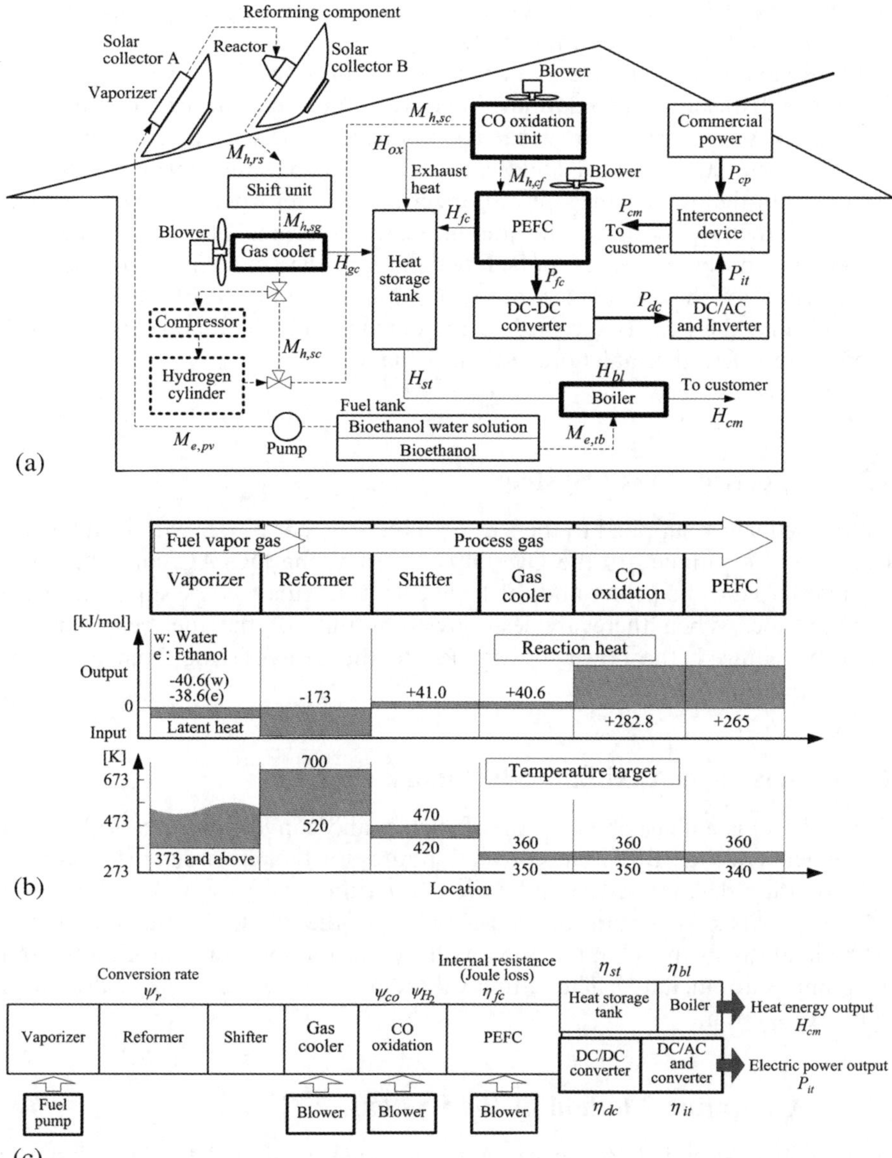

Figure 8.1 System design; (a) Block diagram; (b) Reaction heat and temperature target; (c) Loss of power and heat.

supplied to a reactor installed in solar collector B. The reactor is filled with a reforming catalyst. If the fuel vapor contacts the reforming catalyst, it will be reformed into gas with a high hydrogen content. Surplus water and CO are contained in the reformed gas. Therefore, water is removed using a gas cooler. Moreover, reformed gas is supplied to a shift unit and a CO oxidation unit, and the amount of CO is oxidized. CO included in the compressed hydrogen is also oxidized using the CO oxidation unit.

(2) Reaction temperature and chemical reaction heat.

Figure 8.1(b) shows the relationship between the chemical reaction heat and the goal response temperature in each component of the system. The shifter, the gas cooler, the CO oxidation, and the PEFC are accompanied by an exothermic reaction. On the other hand, the vaporizer and the reformer are accompanied by an endothermic reaction. The amount of endothermals of the vaporizer and the reformer of the FBSR is supplied by the solar collectors. After supplying the exhaust heat of the gas cooler, the CO oxidation, and the PEFC to a heat-storage tank, the heat is supplied to the demand side. The boiler is operated when there is less thermal storage than demand.

8.2.3 Electric Power System

If reformed gas is supplied to the PEFC, direct-current power will be obtained. This power is supplied to the DC–DC converter, the DC–AC converter, and the inverter, and the power and the frequency of regulation are supplied to the demand side. When there are less power outputs of the fuel cell than the quantity required, the power is supplied to the demand side from the commercial power.

8.2.4 Loss and Auxiliary-Machinery Power

Figure 8.1(c) is a chart showing the efficiency and conversion rate of all of the components of the FBSR. The conversion rates for the reforming reaction (ψ_r), CO oxidation reaction (ψ_{co}), and hydrogen burning (ψ_{H_2}) are provided. Section 8.4.2.3 describes the generation efficiency (η_{fc}) of the PEFC. On the other hand, in the analysis in this chapter, the auxiliary-machinery power of blowers and the pump is not included. The setting values of η_{st}, η_{bl}, η_{dc}, and η_{it} are described in a later section.

8.2.5 Operation Method of the System

When fuel is supplied to the vaporizer the fuel gas is fed to the reforming component. Reformed gas is supplied to the shift unit and the gas cooler, and is stored using cylinders. In order to store hydrogen, the storage tank of the reformed gas is installed in the system. While compressing and storing the reformed gas, it removes the water vapor in the gas using a cooler. Reformed gas is fed to the CO oxidation unit from the cylinder or the gas cooler, and is supplied to the fuel cell. To output power to the commercial power grid (with a regular frequency and voltage), the output of the fuel cell is converted with a DC–DC converter and an inverter. The exhaust heat from the gas cooler, the fuel cell, and the CO oxidization unit is stored. This heat is supplied to the demand side of the FBSR. When supplied solar heat is insufficient, a boiler is operated.

8.3 Heat-Transfer Analysis

8.3.1 Efficiency of Reforming Component

The reforming component shown in Figure 8.1(a) consists of solar collector B and the reactor (Figure 8.2(a)). A reactor end face is a solar insolation acceptance surface with area A_{hs}. Other surfaces are insulated. The solar insolation collected by solar collector B is input into the solar insolation acceptance surface. Moreover, the heat input to this solar insolation acceptance surface heats the catalyst layer of the reactor. The efficiency of the reforming component is determined by the ratio of the rate of the higher heating value of the hydrogen produced to that of the amount of condensed solar radiation in solar collectors A and B. Equation (8.1) is the formula for the efficiency η_s of the reforming component.

$$\eta_s = \frac{\text{The higher heating value of hydrogen } (Q_h)}{\text{Amount of heat collections per day } (Q_A + Q_B)} \tag{8.1}$$

8.3.2 Heat Transfer in the Catalyst Layer

The reactor is filled with several millimeters of the spherical reforming catalyst. In addition, some of solar insolation input from the acceptance surface on the reactor is discharged to the ambient air by convection heat transfer q_{con} and radiation heat transfer q_{rad}. Equation (8.2) is the heat convection of the catalyst layer, and it contains the Damkohler correction number.[10] The value of Da in eqn (8.2) is calculated using eqn (8.3),

$$Nu = 9.49 \cdot (Re \cdot Pr)^{0.516} \cdot \left({}^{D_c}/_{D_{cl}}\right)^{1.43} + 27.2 \cdot Da^{0.325} \tag{8.2}$$

$$Da = -(H_r \cdot \alpha_r) \cdot D_c / \left(\rho_g \cdot u_g \cdot C_g \cdot T_g\right) \tag{8.3}$$

8.3.3 Reforming Reaction and Analytical Model for the Catalyst Layer

The experimental result of ethanol steam reforming using the commercial catalyst by Akpan *et al.*, which is shown in Figure 8.2(b), is introduced into the analysis.[11] In this figure, the relationship between the amount of catalyst, the flow rate of ethanol, and the temperature of the catalyst layer and the fuel

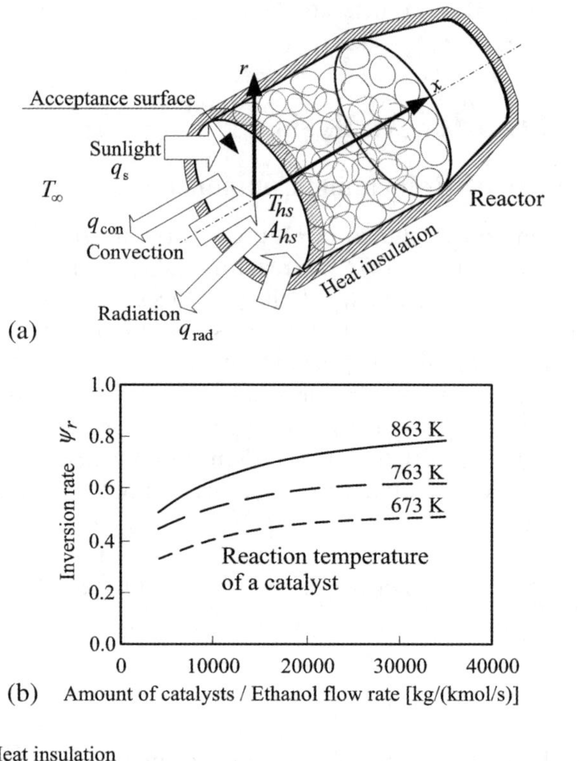

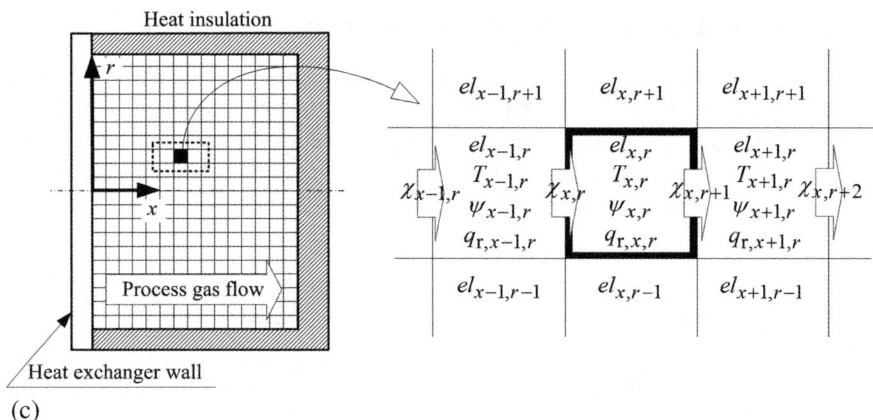

Figure 8.2 Reactor; (a) Reactor; (b) Catalyst performance; (c) Analysis of the heat transfer.

conversion rate of the catalyst is depicted. The fuel conversion rate increases with an increasing amount of catalyst and with increasing temperature, as shown in Figure 8.2(b). Equation (8.4) is the two-dimensional heat diffusion equation of the catalyst layer. The temperature of the catalyst layer is represented by T, the coordinates of the direction of the radial direction is represented by r, and the coordinates of the direction of the catalyst layer is represented by x. A surface element is created in the direction of r, and the direction of x, about the catalyst layer of the cylinder type shown in Figure 8.2(a). Figure 8.2(c) portrays the element analysis model of the catalyst layer. The element number of the catalyst layer is expressed with $el_{x,r}$ (however, $x=1,2,\ldots,N_x$, $r=1,2,\ldots,N_r$).

$$\left(\frac{\partial^2 T}{\partial r^2}+\frac{1}{r}\cdot\frac{\partial T}{\partial r}+\frac{\partial^2 T}{\partial x^2}\right)+q_r=\frac{\rho_c\cdot C_c}{\lambda_c}\cdot\frac{\partial T}{\partial t} \tag{8.4}$$

Boundary conditions

$$\frac{\partial T}{\partial r}=0 \quad \text{at} \quad r=R_{cl}, 0\leq x\leq L_{cl} \tag{8.5}$$

$$\begin{aligned}-\lambda_c\cdot\frac{\partial T}{\partial x}&=q_s-q_{\text{rad}}-q_{\text{con}}\\&=q_s-\varepsilon\cdot\sigma\cdot\left(T_s^4-T_\infty^4\right)-h\cdot\left(T_s-T_\infty\right)\end{aligned} \tag{8.6}$$

at $x=0, \quad 0\leq r\leq R_{cl}$

$$\begin{aligned}&\frac{\partial T}{\partial x}=0 \quad \text{at} \quad x=L_{cl}, \quad \frac{\partial T}{\partial r}=0 \quad \text{at} \quad r=0\\&T=T_\infty \quad \text{for} \quad t=0\end{aligned} \tag{8.7}$$

$$\frac{\partial}{\partial x}\left(\rho_g\cdot u_g\right)=0 \tag{8.8}$$

$$\frac{\partial u}{\partial r}=0 \quad \text{at} \quad r=R_{cl}, \quad \frac{\partial u}{\partial x}=0 \quad \text{at} \quad x=L_{cl} \tag{8.9}$$

$$u=u_0 \quad \text{at} \quad x=0, \quad \frac{\partial u}{\partial r}=0 \quad \text{at} \quad r=0 \tag{8.10}$$

$$q_r=g_g\cdot\psi_r\cdot H_r \tag{8.11}$$

8.3.4 Heat Diffusion Equation

Equation (8.4) is discretized, and the temperature distribution of the catalyst layer is analyzed under the boundary conditions of eqns (8.5)–(8.7). The central finite difference method[12] is used to calculate the discretize eqn (8.4) in this chapter. Here, q_s is the heat concerning the heat exchange wall of the reactor. Equations (8.5)–(8.11) are given as the boundary conditions when calculating eqn (8.4). Equation (8.7) is the boundary condition of temperature and eqn (8.8) is the boundary condition of mass flow of process gas. Equations (8.9) and (8.10) are the boundary conditions of space velocity. Furthermore, eqn (8.11) is the boundary condition of molar flow rate of the process gas. Equation (8.8) is the mass flow rate of the process gas, and eqns (8.9) and (8.10) are the boundary conditions in this analysis. The volume flow rate of the process gas is u_g in eqn (8.8), and ρ_g is the mean density. The value u_0 in eqn (8.10) is the space velocity of the fuel vapor at the entrance of the catalyst layer. This value is the result of dividing the volume flow rate of the fuel vapor by the cross section of the catalyst layer. eqn (8.11) expresses endothermals from the reforming reaction. Variables g_g, ψ, and H_r in eqn (8.11) express the molar flow rate of process gas, the conversion ratio, and the reaction heat, respectively. If the temperature T of the catalyst layer is known, the conversion rate ψ can be obtained from the characteristic of the catalyst. Because H_r is determined by the reforming reaction, if g_g is given, the amount of endothermals q_r can be calculated using the reforming reaction.

8.4 Analysis Method

8.4.1 Temperature Distribution of the Catalyst Layer, and the Composition Distribution

The temperature distribution ($T_{x,r}$) of the catalyst layer shown in Figure 8.2(c) is obtained by introducing calculus of finite differences into eqn (8.4). The boundary conditions at this time are shown in eqns (8.5)–(8.10). The Gauss–Seidel method is used for convergence calculation of the calculus of finite differences. If the temperature distribution $T_{x,r}$ is decided, the conversion rate ($\psi_{x,r}$) in each element is obtained based on the relationship between catalyst temperature and conversion rate (Figure 8.2(b)). The amount of endothermals of process gas ($q_{r,x,r}$) is calculated based on the value of $\psi_{x,r}$. The temperature distribution ($T_{x,r}$) of the catalyst layer is calculated from these results, and this calculation is repeated to convergence of the discretization equation (of eqn (8.4).

8.4.2 Amount of Exhaust Heat

8.4.2.1 Gas Cooler

The exhaust heat of the gas cooler is calculated using eqn (8.12) where the number of components in the process gas is N_g and Δt is the difference in temperature between the process gas in the gas cooler entrance and exit. C_n and G_n are the specific heat and mass flow rates of the gas composition n,

$$H_{gc} = \Delta t \cdot \sum_{n=1}^{N_g} (C_n \cdot G_n) \tag{8.12}$$

8.4.2.2 CO Oxidation Unit

In the CO oxidation unit, the CO concentration is reduced by CO burning. However, the hydrogen contained in the process gas in the case of CO burning may also burn. This selectivity is decided by control of the catalyst temperature. In reality, some of the hydrogen also burns under strict temperature control. In this chapter, the rate of hydrogen burning in the process gas is set at 3%. The exhaust heat of the CO oxidation unit is calculated using eqn (8.13). Here, V is the volume flow rate of each gas and J is the calorific value.

$$H_{\mathrm{ox}} = V_{\mathrm{CO}} \cdot J_{\mathrm{CO}} + 0.03 \cdot V_{\mathrm{H_2}} \cdot J_{\mathrm{H_2}} \tag{8.13}$$

8.4.2.3 PEFC

When a PEFC of small capacity compared to the power demand amount is introduced, the operation time of a full load is long. Generally, the exhaust heat power of a PEFC differs with load factor. The PEFC in this chapter has many full-load operation hours because the capacity is small compared to the power demand. The PEFC often has to operate with a high load factor. Therefore, the generation efficiency (η_{fc}) of the PEFC is set as 65%. On the other hand, the exhaust heat of the PEFC is calculated to be 30% of the calorific power of supply hydrogen.

8.5 Operation Case

8.5.1 Specification of the Reforming Component

(1) Catalyst layer.

The FBSR was introduced into individual houses in Sapporo, Japan. The specification of the reforming component introduced into this case analysis is described in Table 8.1. The diameter of the catalyst layer D_{cl} is 80 mm and its

Table 8.1 Analysis conditions.

Each concentration area of the solar collector A and B	1.0 m^2
Reactor	
Length of the catalyst layer (L_{cl})	60 mm
Diameter of the catalyst layer (D_{cl})	80 mm
Panicle diameter of the catalyst (D_c)	3.0 mm
Steam/carbon ratio	3.0
Catalyst filling factor	0.85
Sampling time	0.01 s
The number of element of.v-axis (N_x)	30
The number of element of r-axis (N_r)	40
Density of the catalyst	213kg/m^5
Heat conductivity of the catalyst	lOW/m K
Efficiency	
DC–DC converter (η_{dc})	95%
DC–AC converter and inverter (η_{it})	95%
Boiler (η_{bl})	90%
Heat storage (η_{st})	90%
Loss of the CO oxidation unit	3%

width L_{cl} is 60 mm. The reactor is filled with the reforming catalyst of 3 mm average particle diameter. The catalyst filling factor is 0.85. The detailed experimental results of the ethanol steam reforming by this reforming catalyst are described by reference 6. In addition, the catalyst layer is divided into elements of 2 mm length in the directions of r and x. The numbers of elements are $N_x = 30$ and $N_r = 40$.

(2) Heat system.

The heat-transfer coefficient h_∞ of the convective heat transfer q_{con} described by Section 8.3.2 was set at 10 W/m^2 K (natural convection). Moreover, ε_{hs} of the radiation heat transfer q_{rad} gives 0.95 assuming a black body. The area A_{hs} on the solar insolation acceptance surface of the reactor is 0.005 m^2. The transmissivity of the heat exchange wall of the reactor was set at 0.9, and the condensing efficiency of the solar collector was set at 90%.

(3) Solar collector and fuel supply.

In the analysis in this chapter, the collecting area of solar collector A and solar collector B were both set to be 1 m^2 (unit area). In the convergence calculation of the discretization equation, an analytical accuracy of 10^{-5} was used. The use of an ethanol solution fuel supply determined that the value of the horizontal axis in Figure 8.2(b) (amount of catalysts /ethanol flow rate) is 35 000 kg/(kmol/s).

8.5.2 Storage of the Reformed Gas

The reformed gas can be compressed and stored in the FBSR (broken-line block in Figure 8.1(a)). Accordingly, the reformed gas with the pressure P_0 and the flow rate $U_{0,t}$ output from the gas cooler is pressurized to P_{cp} with a compressor. Here, the subscript t is sampling time. The work $W_{p,t}$ of the compressor is assumed to be the work of compression by an ideal gas and is calculated using eqn (8.14). The compressor efficiency η_{cp} in the equation includes the electricity consumption in an electric motor, the transmission loss of power, loss due to insufficient air leak and cooling, and other mechanical loss.

$$W_{p,t} = P_0 U_{0,t} \ln\left(P_{cp}/P_0\right) \Big/ \eta_{cp} \tag{8.14}$$

8.5.3 Installation Requirements of the System and Demand Characteristic

The operation of the power and heat in the case of introducing the FBSR into standard houses in Sapporo is planned. However, since a detailed report about the operation method of this power has already been published,[8] this chapter focuses on examining the overall efficiency of the system.

The power load and the heat consumption of a house in Sapporo on a representative day in both March (March 1) and August (August 23) are shown in Figues 3(a) and (b).[13] Because Sapporo is located in a cold district, there is no cooling load in August. On the other hand, the space heating load in March is supplied from the exhaust heat of the system and the backup boiler. As shown in Figure 8.3, the characteristics concerning the amount of solar radiation and outside air temperature[14] will differ greatly in March and August. In this chapter, as observational data of the solar irradiance and the outside air temperature, Surface-Weather-Observation 1–Minute Data and 2007 Sapporo District Meteorological Observatory and Japan Meteorological Business Support Center[14] are used. The figure at the bottom of Figure 8.3 depicts daylight hours on each representative day. The amount of solar radiation, outside air temperature, and daylight hours are data collected at 1–min intervals.

8.6 Results and Discussion

8.6.1 Temperature Distribution of the Catalyst Layer

Figure 8.4 shows the result of the transient response characteristic of the catalyst layer temperature when inputting constant solar insolation (250 W/m^2, 500 W/m^2, 1000 W/m^2) into the reactor.[8] In Figure 8.4, 0 s is the stable time of the acceptance surface temperature after inputting solar insolation into the reactor. When the outside air temperature is 293 K, the acceptance surface with solar irradiance of 250 W/m^2 rises to about 500 K. In the input of 1000 W/m^2, the acceptance surface rises to about 890 K. The conversion rate of ethanol steam reforming increases, which leads to a high catalyst temperature in the reactor. Therefore, the temperature distribution of the catalyst layer shown in Figure 8.4 differs so greatly that solar irradiance is large.

8.6.2 Composition of the Process Gas

Figure 8.5 shows the analysis result of the process gas composition of the catalyst layer when inputting constant solar insolation (250 W/m^2 and 500 W/m^2) into the reactor.[8] This figure shows the process gas composition along the x-axis of the catalyst layer as predicted by the analysis. This figure shows the composition of each gas with its respective molar flow rate. The molar flow rate of hydrogen is larger than other gases in the composition. Distribution of the molar flow rate of hydrogen, and the time at which the hydrogen flow rate becomes stable are influenced by the magnitude of the solar irradiance input into the reactor. If there is little solar irradiance and there is a short period of solar radiation fluctuation, the hydrogen generation rate may not reach the maximum possible. From this result, when there is short-term fluctuation solar radiation with little solar irradiance, the hydrogen generating rate may not reach a stable generation rate (rated speed) by a response delay.

8.6.3 Amount of Hydrogen Generated

Figure 8.6 shows the analysis result of the amount of hydrogen generated by the FBSR using the amount of solar radiation, outside air temperature, and

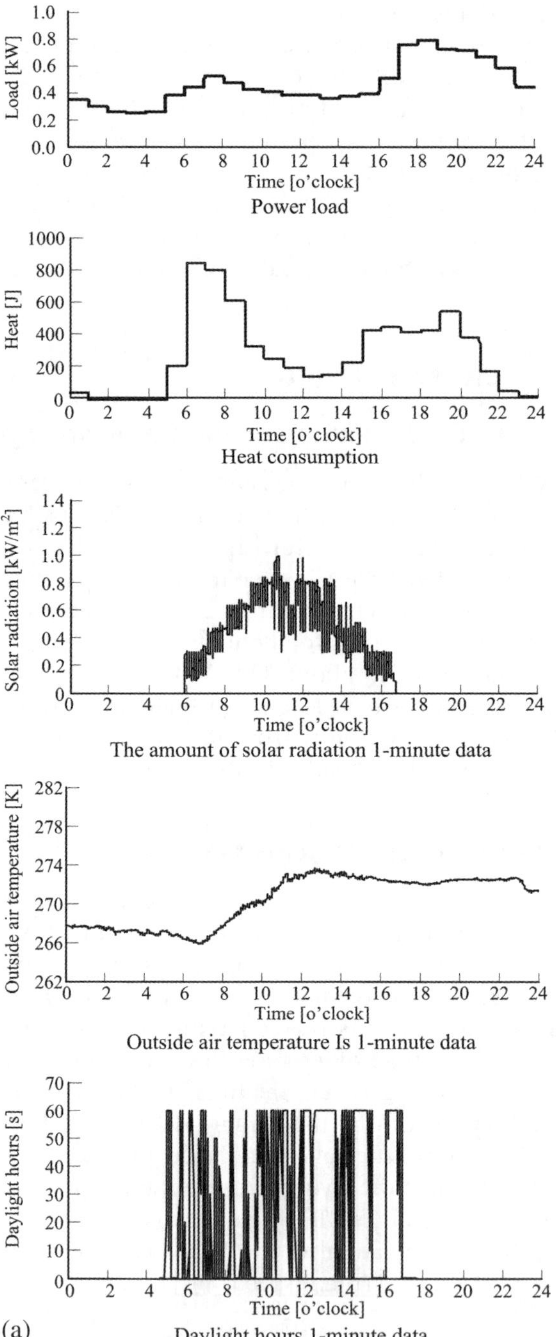

Figure 8.3 Weather observation 1–min data in Sapporo; (a) March 1, 2007; (b) August 23, 2007.

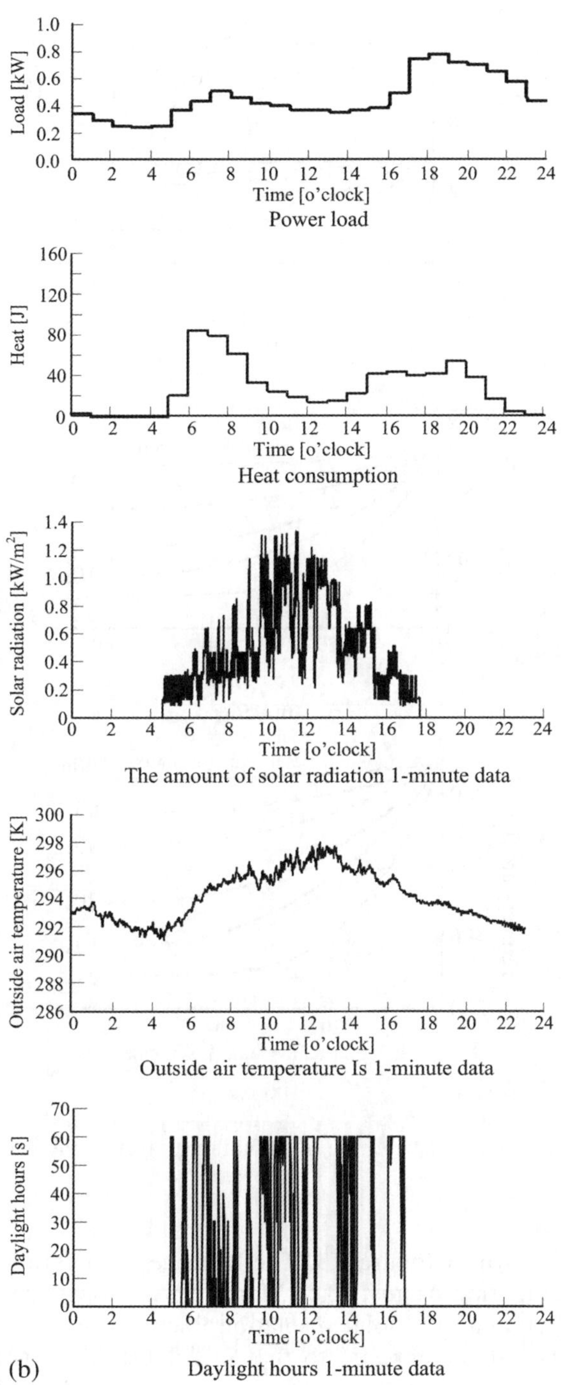

Figure 8.3 Continued.

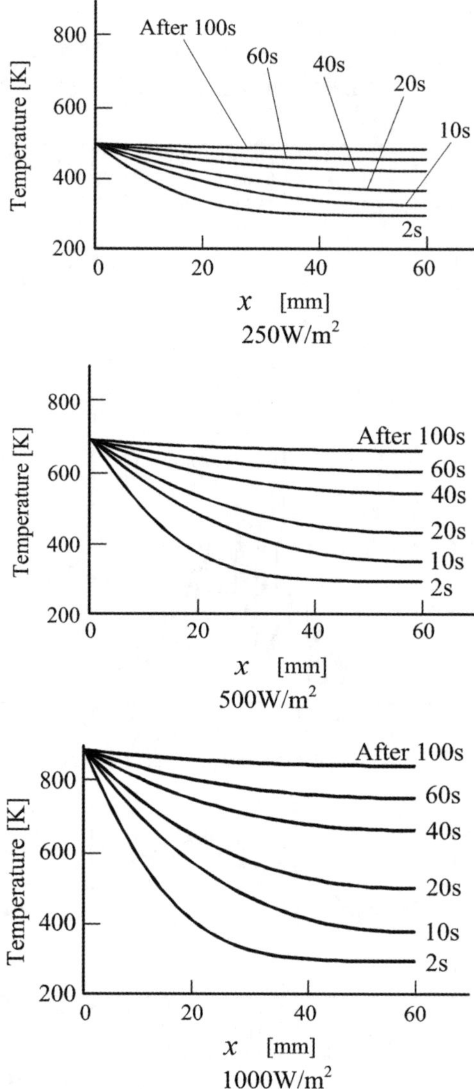

Figure 8.4 Temperature distribution in the catalyst layer of the reactor.

daylight hours shown in Figure 8.3. The hydrogen generating rate fluctuates greatly when comparing August 23 to March 1. As shown in Figure 8.3, from 6:00 a.m. to 11:00 a.m., the solar insolation fluctuation on the representative day in March is stable compared to that in August. The cause of this solar insolation fluctuation is shadowing by clouds. As a result, in Figure 8.6, the amount of hydrogen production on a representative morning in March will be stabilized compared with that in August. As shown in Figure 8.4, the rate of

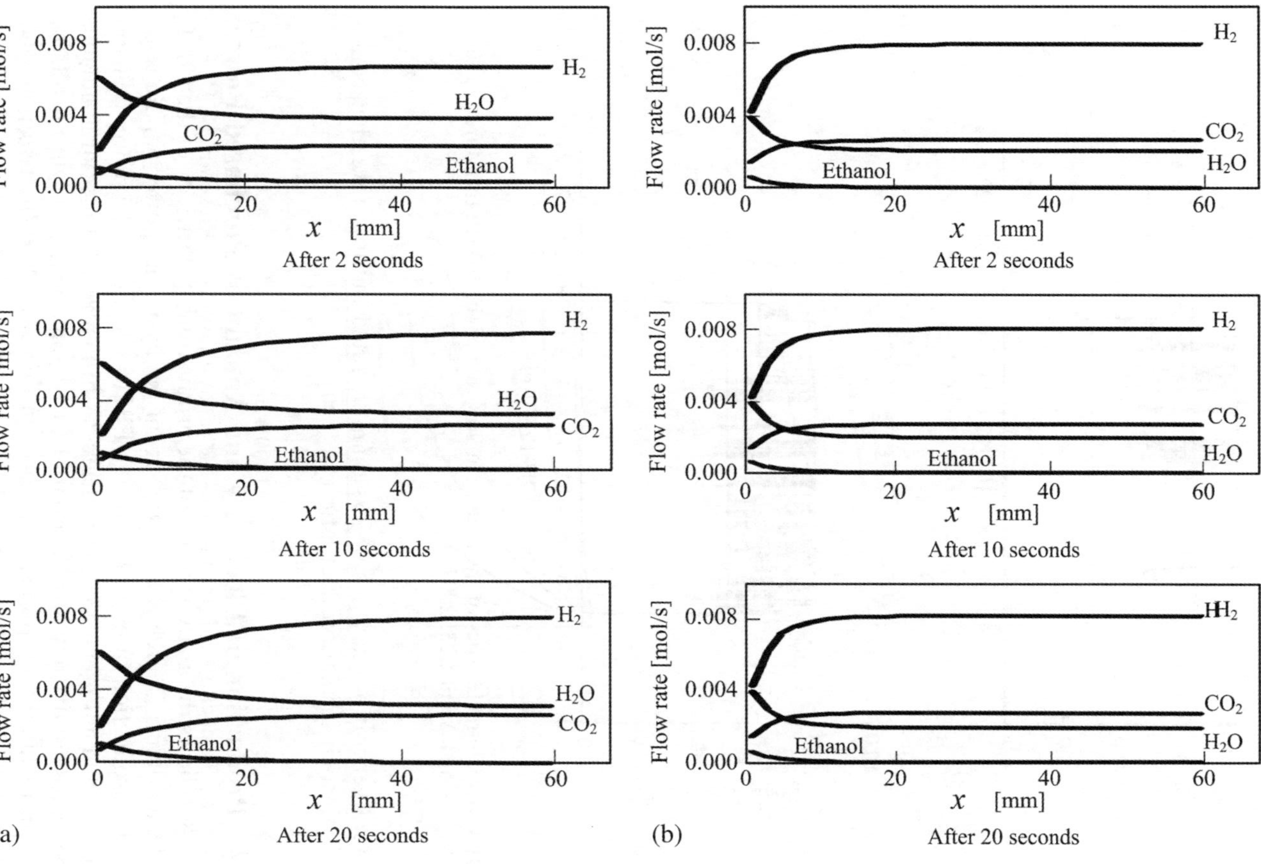

Figure 8.5 Flow rate of process gas in the catalyst layer. Outside air temperature is 293 K. (a) The amount of heat collection of solar collector B is 250 W/m2; (b) The amount of heat collection of solar collector B is 500 W/m2.

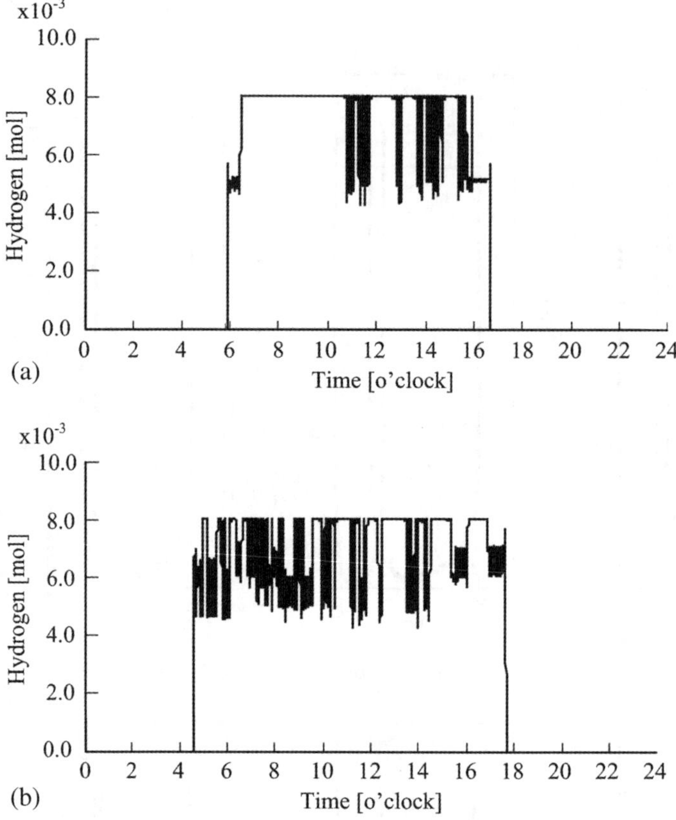

Figure 8.6 Characteristics of the hydrogen flow rate of the FBSR; (a) March 1, 2007; (b) August 23, 2007.

hydrogen generation may not be less than the rated output in weather conditions with solar insolation fluctuation.

8.6.4 Production of Electricity and Amount of Purchased Power

Figure 8.7 shows the analysis result of the production of electricity of the FBSR and the amount of purchased power. In this analysis, the power is purchased when the production of electricity of the FBSR is less than the power demand amount shown in Figure 8.3. As for the result of Figure 8.7, the solar irradiance on the representative day of each month and the characteristic of the outside air temperature affect the purchase power. Moreover, when introducing solar collector A and solar collector B with areas of 1.0 m^2, the power load peak at 8:00 of both representative days (Figure 8.3) can be decreased. However, in order to decrease the peak around 19:00, the increase in the compressed hydrogen by extension of the solar collecting area is required. Moreover, a time

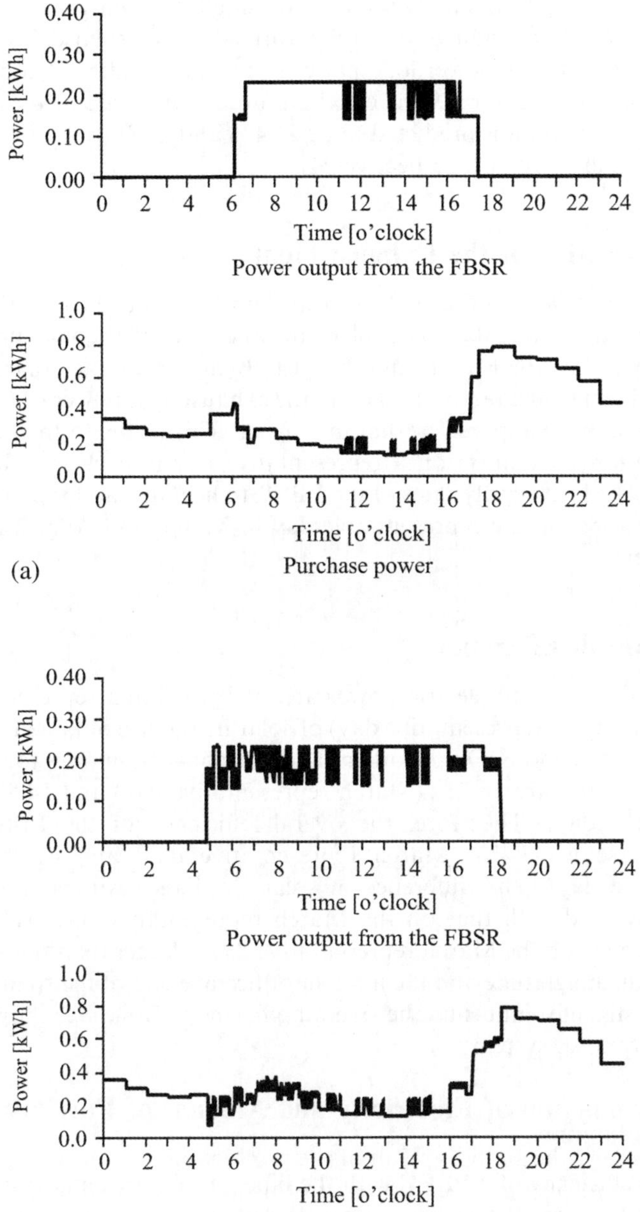

Figure 8.7 Analysis results of the power operation plan; (a) March 1, 2007; (b) August 23, 2007.

shift of the electric power supply is required using hydrogen storage equipment. When the amount of production of the reformed gas by solar collectors A and B is introduced into the power load pattern shown in Figure 8.3, storage of the reformed gas is not required. The electric power supply rate of the FBSR to a power demand amount is predicted to be 21.4% and 25.3% on a representative day in March and August, respectively.

8.6.5 Operation of the Exhaust Heat

Figures 8.8 and 8.9 show the analysis result of the exhaust heat of the fuel cell, the CO oxidation unit, the gas cooler, and the heat of the backup boiler on representative days in March and August. In terms of the solar insolation fluctuation from 6:00 a.m. to 11:00 a.m., the exhaust heat power of the morning in March is stable compared to that in August. This is due to the difference in solar insolation fluctuation on a representative day in each month shown in Figure 8.3. The heat supply rates of the FBSR to heat demand is predicted to be 1.2% and 13.7% on the representative day in March and August (except for boiler power).

8.6.6 Overall Efficiency

The conversion rates to electric power are 30.7% (March representative day) and 27.1% (August representative day) of solar irradiance obtained by the 1 m^2 solar collectors A and B. On the other hand, the conversion rates to heat supply of solar irradiance are 16.7% (March representative day) and 14.8% (August representative day). Therefore, the overall efficiency of the FBSR by this operation case is 47.4% (March representative day) and 41.9% (August representative day). The difference in solar irradiance will be 1.32 times in August compared with that on the March representative day. However, the overall efficiency on the March representative day is larger than that in August. Therefore, the magnitude and the number of occurrences of the solar insolation fluctuation strongly influence the overall efficiency. Table 8.2 shows analysis results of proposal system

8.7 Conclusions

The overall efficiency of a PEFC with the bioethanol reforming system using a sunlight heat source (FBSR) was investigated by numerical analysis. In this chapter, the heat transmission characteristics of the catalyst layer installed in the reactor were investigated. The transient characteristic of hydrogen generation was examined based on these results. Furthermore, the supply and amount of purchase of electric power and heat were investigated using the energy–demand characteristic in a standard house in addition to meteorological data on representative days in March and August in Sapporo, Japan. The total collecting area of the solar collectors was 2 m^2. The following

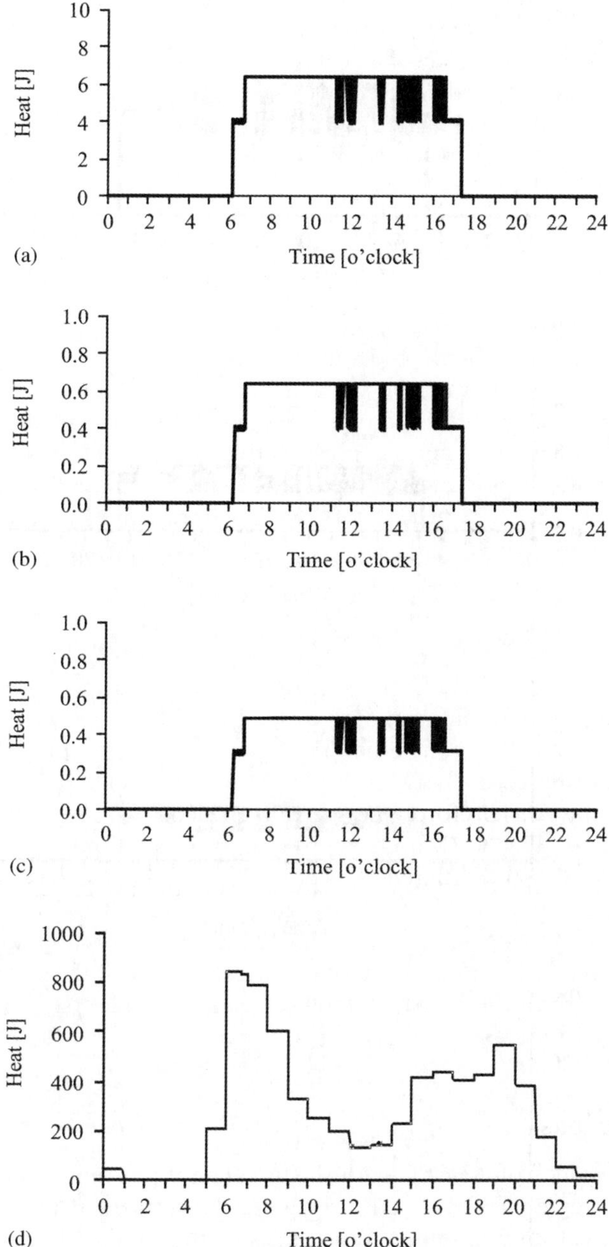

Figure 8.8 Analysis results of the heat operation plan on March 1, 2007. (a) Exhaust heat of the FC in one-minute intervals; (b) Exhaust heat of the CO oxidation unit in one-minute intervals; (c) Exhaust heat of the gas cooler in one-minute intervals; (d) Purchased heat in one-minute intervals.

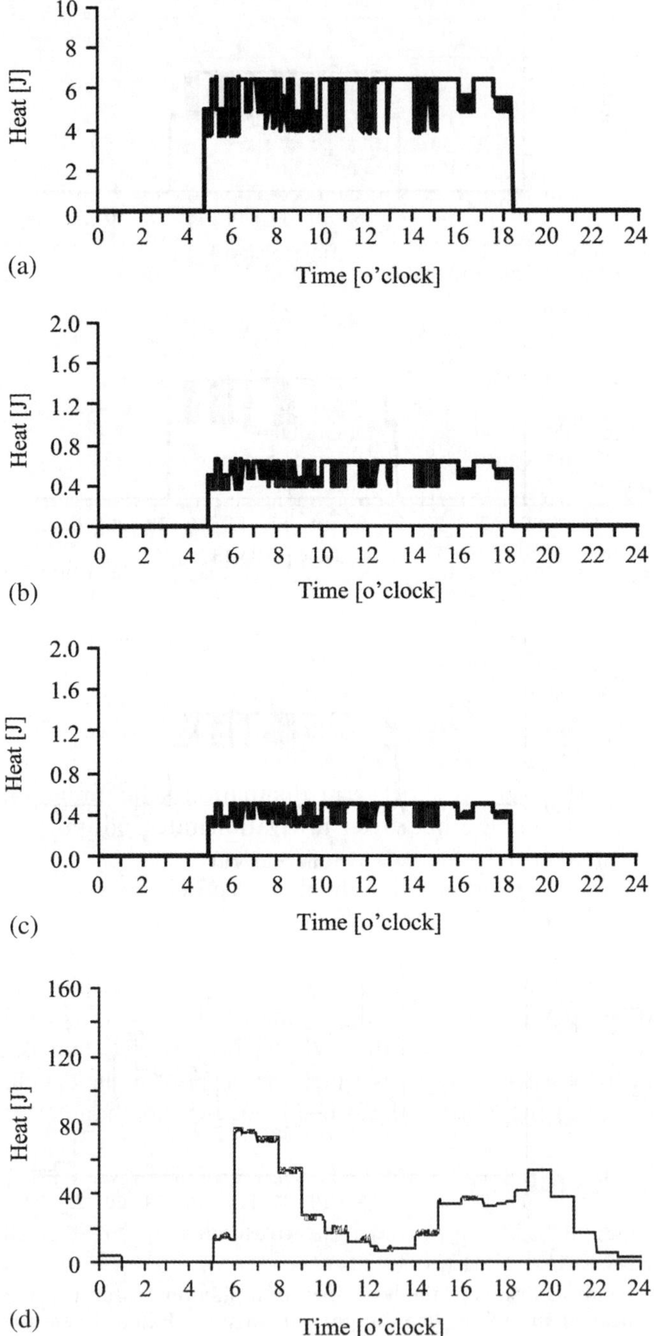

Figure 8.9 Analysis results of the Heat Operation plan on August 23, 2007. (a) Exhaust heat of the FC in one-minute intervals; (b) Exhaust heat of the CO oxidation unit in one-minute intervals; (c) Exhaust heat of the gas cooler in one-minute intervals; (d) Purchased heat in one-minute intervals.

Table 8.2 Analysis results of the FBSR performance.

	March 1	*August 23*
Amount of solar radiation per (Jay by the solar collectors A and B	28.0 MJ/day	37.0 MJ/day
Amount of hydrogen production per day	100 g/day	117 g/day
Efficiency of a reforming component (The higher calorific value of hydrogen/amount of heat collections per day)	47%	42%
Amount of power demand per day	11.16 kWh	11.03 kWh
Amount of power generation per day	2.39 kWh	2.79 kWh
Amount of CO; emissions per day	732 g/day	854 g/day
Use rate of renewable energy (Condensing area 2.0 m^2)		
Power	30.7%	27.1%
Heat	16.7%	14.8%
Total	**47.4%**	**41.9%**
System output to the quantity power demanded	21.4%	25.3%
System output to the quantity heat demanded (except for the boiler)	1.2%	13.7%

conclusions were drawn from these analysis results. As uncertainty in this analysis, the calculation error, the setting performance of each equipment, *etc.* can be considered. The magnitude of these influences and error concerning the grid system will be explained by future study.

(1) The magnitude of solar irradiance greatly influences the temperature distribution and composition distribution of process gas in the catalyst layer. When there is short-term fluctuation solar radiation with little solar irradiance, the hydrogen generation rate may not reach a stable value (rated speed) due to a response delay.

(2) The rate of converting sunlight into electrical power in the proposed system is 30.7% and 27.1% on representative days in March and August, respectively. On the other hand, the rate converted into heat is 16.7% and 14.8%, respectively. As a result, the overall efficiency of the FBSR by the analysis case in this chapter is 47.4% and 41.9%, respectively. These results indicate that the proposed system is competitive with other energy systems, such as a photovoltaic cell. However, operation of the FBSR takes the cost of the bioethanol.

Acknowledgement

This work was partially supported by a Grant-in-Aid for Scientific Research (C) from JSPS.KAKENHI (20560204).

Nomenclature

C: specific heat J/(g K)
Da: Modified Damkohler number

el: element number of the catalyst layer
G: mass flow rate g/s
g_g: molar flow rate mol/s
H: reaction heat J/mol
h: heat-transfer coefficient W/(m^2 h K)
J: calorific value J/m^3
L: length, width m
N: the number of elements
N_g: the number of gas composition
N_u: Nusselt number
P: power W
P_0: inlet pressure of the process gas Pa
P_{cp}: outlet pressure of the process gas Pa
Pr: Prandtl number
Q: quantity of heat J
Q: heat W
R: radius m
r: radial direction of the catalyst layer
Re: Reynolds number
T: temperature K
t: sampling time s
Δt: temperature difference K
U_0: the volume flow rate of the process gas m^3/s
U: flow rate m/s
U_0: initial flow rate m/s
W_p: work of the compressor W
X: axial direction of the catalyst layer

Greek Symbols

α_r: reaction rate mol/(m^3 s)
χ: layer of the element
ε: emissivity
η: efficiency
η_{cp}: efficiency of the compressor
η_s: efficiency of the reforming component
λ: heat conductivity W/(m K)
ρ: density g/m^3
σ: Stefan–Boltzmann constant
ψ: conversion ratio

Subscripts

A, B: solar collectors A and B
bl: boiler
c: catalyst

cl: catalyst layer
con: convective heat transfer
cf: CO oxidation unit to the cell stack
cm: customer
con: convection
cp: commercial power
dc: DC–DC converter
fc: cell stack
g: process gas
gc: the gas cooler to the heat-storage tank
h: the higher heating value of hydrogen
hs: heat supply surface of the reactor
it: DC–AC converter and inverter
ox: the CO oxidation unit
pv: the pump to the vaporizer
r: reforming
rad: radiation
rs: the reactor to the shift unit
s: sunlight
sc: the gas cooler to the CO oxidation unit
st: storage tank
sg: the shift unit to the gas cooler
tb: the fuel tank to the boiler
∞: ambient air

References

1. Shin'ya Obara and Itaru. Tanno, Development of Distributed Energy System due to Bio-ethanol PEM Fuel Cell with Solar Reforming, Part 1—Evaluation of Basic Performance, *Trans. Soc. Heat. Air-Con. Sanitary Eng. Jpn.*, 2007, **123**, 23–32.
2. Shin'ya Obara and Itaru Tanno, Development of Distributed Energy System due to Bio-ethanol PEM Fuel Cell with Solar Reforming, Part 2—High-speed analysis of the operation plan using a neural network, *Trans. Soc. Heat. Air-Con. Sanitary Eng. Jpn.*, 2008, **130**, 33–42.
3. Shin'ya Obara and Itaru Tanno, Operation Prediction of a Bioethanol Solar Reforming System Using a Neural Network, *J. Thermal Sci. Technol.*, 2007, **2**(2), 256–267.
4. Abdul-Majeed Azad, Sathees Kesavan and Sirhan Al-Batty, A Closed-Loop Proposal for Hydrogen Generation Using Steel Waste and a Prototype Solar Concentrator, *Int. J. Energy Res.*, 2009, **33**(5), 481–498.
5. Adam Noglik, Martin. Roeb, Christian. Sattler and Robert. Pitz-Paal, Experimental Study on Sulfur Trioxide Decomposition in a Volumetric Solar Receiver-reactor, *Int. J. Energy Res.*, 2009, **33**(9), 799–812.

6. John Turner, George Sverdrup, Margaret K. Mann, Pin-Ching Maness, Ben. Kroposki, Maria Ghirardi, Robert J. Evans and Dan Blake, Renewable Hydrogen Production, *Int. J. Energy Res.*, 2008, **32**(5), 379–407.
7. Patrice Charvin, Abanades Ste′phane, Lemort Florent and Flamant Gilles, Analysis of solar chemical processes for hydrogen production from water splitting thermochemical cycles, *Energy Conver. Manag.*, 2008, **49**, 1547–1556.
8. Fernando Fresno, Rocío. Fernández-Saavedra, M. Belén Gómez-Mancebo, Alfonso Vidal, Miguel Sánchez, M. Isabel Rucandio, Alberto J. Quejido and Manuel Romero, Solar hydrogen production by two-step thermochemical cycles: Evaluation of the activity of commercial ferrites, *Int. J. Hydrogen Energy*, 2009, **34**(7), 2918–2924.
9. Jin Xuan, Michael K. H. Leung, Dennis Y. C. Leung and Meng Ni, A review of biomass-derived fuel processors for fuel cell systems, *Renew. Sustain. Energy Rev.*, 2009, **13**(6–7), 1301–1313.
10. Y. Usami, S. Fukusako and M. Yamada, Heat and Mass Transfer in a Reforming Catalyst Bed (Quantitative Evaluation of the Controlling Factor by Experiment), *Trans. JSME, Series B*, 2000, **67**(659), 1801–1808.
11. E. Akpan, A. Akande, A. Aboudheir, H. Ibrahim and R. Idem, Experimental, Kinetic and 2-D Reactor Modeling for Simulation of the Production of Hydrogen by the Catalytic Reforming of Concentrated Crude Ethanol (CRCCE) Over a Ni-Based Commercial Catalyst in a Packed-Bed Tubular Reactor, *Chem. Eng. Sci.*, 2007, **62**(12), 3112–3126.
12. Mitchell and Andrew Ronald, *Finite Difference and Related Methods for Differential Equations,* Wiley, New York, 2001.
13. K. Narita, *The Research on Unused Energy of the Cold Region City and Utilization for the District Heat and Cooling*, Ph.D. thesis, 1996, Hokkaido University.
14. *Surface-weather-observation 1-minute data, 2007 Sapporo district meteorological observatory*, Japan Meteorological Business Support Center, 2008, Tokyo.

CHAPTER 9

PEFC/Green Energy Compound System (3) – Fuel Cell Microgrid with Wind-Power Generation

SHIN'YA OBARA

9.1 Introduction

It is predicted that a microgrid technique is effective as a backup power supply in an emergency, a peak cut of power plants, and exhaust heat utilization. Furthermore, when renewable energy is connected to a microgrid, there is potential to reduce the amount of greenhouse-gas discharge.[1–3] A microgrid has two varieties, a system interconnection type and an independent type. The microgrid with an interconnection system outputs and inputs the power between other grids. Therefore, the dynamic characteristic of the grid is influenced by the grid at the connection destination. When a microgrid and a large-scale grid such as a commercial power system are interconnected, the dynamic characteristics of the power depend on the commercial power system. For this reason, in the microgrid of the interconnection type, the option of the equipment to connect is wide. On the other hand, since a microgrid can reduce transportation loss of power and heat, this technique may become the major energy supply. The method of connecting two or more small-scale fuel cells and renewable energy equipment by a microgrid, and supplying power to the demand side is effective in respect of environmental problems. So, this chapter examines the independent microgrid that connects fuel cells and wind-power

RSC Energy Series No. 3
Compound Energy Systems: Optimal Operation Methods
By Shin'ya Obara and Arif Hepbasli

Published by the Royal Society of Chemistry, www.rsc.org

generation. In order to follow load fluctuation with an independent grid system, there is a method of installing a battery and a method of controlling the output of power generators. Since the battery is expensive, in this chapter, it corresponds to load fluctuation by controlling the power output of the fuel cell. The output adjustment of the fuel cell has the method of controlling the production of electricity of each fuel cell, and the method of controlling the number of operations of the fuel cell. However, adjustment of the production of electricity of each fuel cell connected to the microgrid may operate some fuel cells with a partial load with low efficiency. So, in this chapter, the number of operations of fuel cells is controlled to follow fluctuations in the electricity demand.

In an independent microgrid, a certain fuel cell connected to the microgrid is chosen, and it is considered as a power basis. The power (voltage and frequency) of the other fuel cells is controlled to synchronize with this base power. Therefore, if the fuel cell that outputs base power is unstable, the power quality of the whole grid will deteriorate. Fuel cells other than base load operation are controlled to synchronize with the base power. The power quality (voltage and frequency) of the microgrid depends on the difference in the demand-and-supply balance.

A 2.5-kW fuel cell is installed in one house of the microgrid formed from ten houses. This fuel cell is operated corresponding to a base load. A 1-kW fuel cell is installed in seven houses, and a 1.5-kW wind power generator is connected to the microgrid. According to the difference in electricity demand of the grid and power produced by the wind power generator, the number of operations of 1-kW fuel cells is controlled. A city-gas reformer is installed in houses in which fuel cells are installed, and hydrogen is produced by city gas reforming. By adding random fluctuation to an average power load pattern, the power demand of a general residence is simulated and is used for analysis. The dynamic characteristics of the microgrid and the efficiency of the system that are assumed in this chapter are investigated by numerical analysis.

9.2 Microgrid Model

Figure 9.1 shows the fuel-cell-independent microgrid model investigated in this chapter. There is a network of power and city gas in this microgrid. Although a power network connects all houses, a city gas network connects houses in which a fuel cell is installed. The fuel cell installed in each house is a proton-exchange membrane type (PEFC). The output of a 2.5-kW fuel cell is fixed to be a base power of the microgrid. Moreover, PEFC of 1 kW power is installed in seven houses. However, the fundamental dynamic characteristics of all the fuel cells are the same, and a fuel cell and a city-gas reformer are installed as a pair. One set of wind power generators is installed, and the power produced by wind force is supplied to a microgrid through an inverter and an interconnection device. The power supply of the microgrid assumes 50 Hz of the single-phase 200 V.

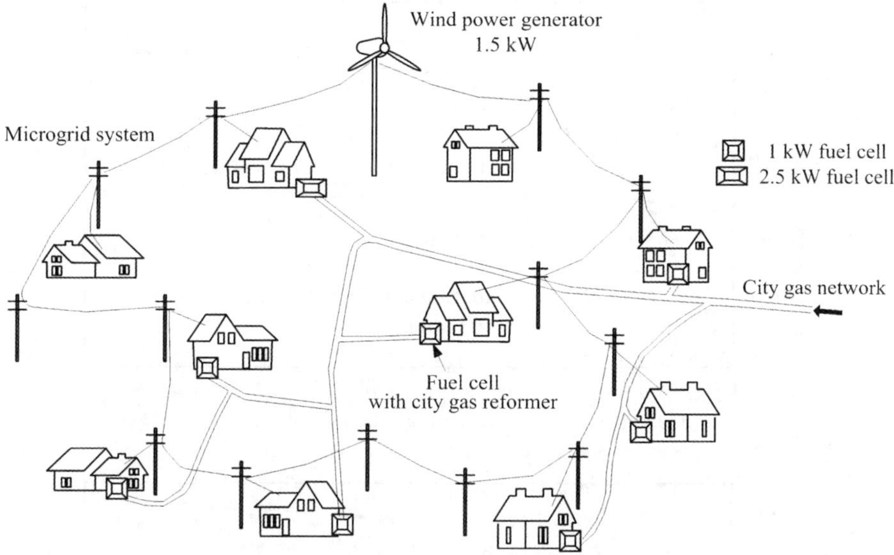

Figure 9.1 Fuel cell microgrid system with wind power generator.

9.3 Response Characteristic of System Configuration Equipment

9.3.1 Power-Generation Characteristic of Fuel Cell

Figure 9.2(a) shows the result of measurement when inputting a load of 45 W into the testing equipment of PEFC (maximum output 100 W) stepwise. In the test, the ambient temperature was set to 293 K, and reformed gas and air were supplied to an anode and a cathode, respectively. An approximated curve is prepared from the result of the measurement in Figure 9.2(a), and the transfer function of a primary delay is obtained. Strictly, although a transfer function is considered depending on the load factor, it is not taken into consideration because this difference is small by test results.

9.3.2 Output Characteristics of City Gas Reformer

Figure 9.2(b) shows the output model that input a load of 100% load factor into the city-gas reformer stepwise.[4–9] An approximated curve is prepared from the result of the measurement, and the transfer function of the primary delay of the city-gas reformer is obtained. As a fuel cell, although the transfer function of a city-gas reformer influences the magnitude of the load significantly, since there is no large difference, the result of Figure 9.2(b) is used. Compared with the condition of the steady operation of the reformer, the characteristics of a startup and a shutdown differ greatly. Cold-start operation and shutdown

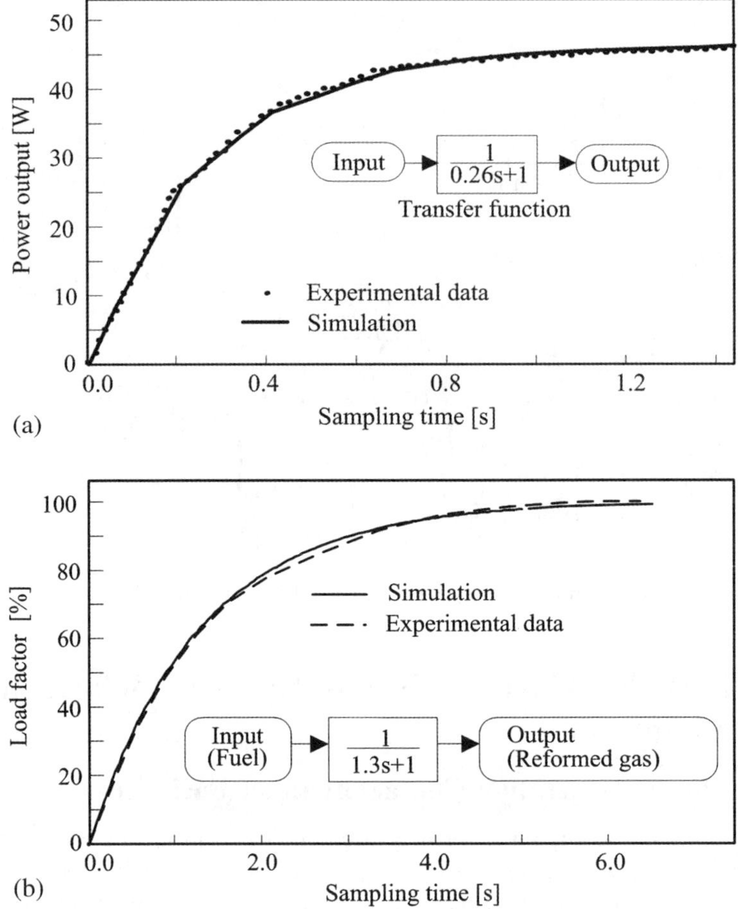

Figure 9.2 Response characteristics of system configuration equipment; (a) Characteristics of transient response for a PEFC; (b) Characteristics of transient response of a reformer.

operation require about 20 min, respectively. In the analysis of this chapter, it is assumed that the startup of the methanol reformer is always a hot start.

9.3.3 Power-Generation Characteristics of Wind-Power Generation

The model of power obtained by wind-power generation is decided at random between 0 to 1.5 kW for every sampling time, as shown in Figure 9.3(a). The power of the wind power generator is supplied to a microgrid through an inverter and a system-interconnection device. Figure 9.3(b) shows the output model of the wind power generator through an inverter and a system-interconnection device. Because the influence is taken in the dynamic

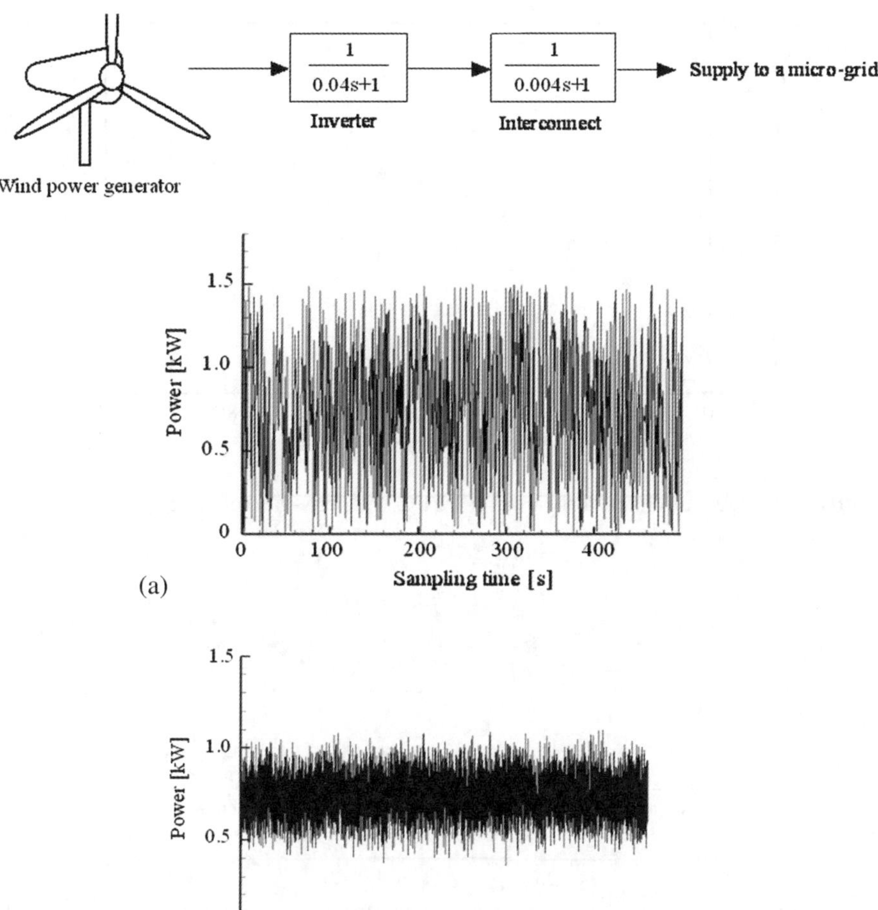

Figure 9.3 Output model of wind power generator; (a) Output of wind power generator; (b) Power supply to a microgrid.

characteristic of an inverter and a system-interconnection device, the output of wind-power generation is settled on a width of 0.75 ± 0.25 kW, as shown in Figure 9.3(b). The details of the transfer function of an inverter and a system-interconnection device are given with Section 9.3.5. The dynamic characteristics of the inverter and system-interconnection device significantly influence the power output characteristics of wind-power generation.

9.3.4 Generation Efficiency of the Fuel Cell System

Figure 9.4 shows a model of the relation between the load factor of a fuel cell, and generation efficiency.[5,10] Power-generation efficiency is obtained by dividing "the

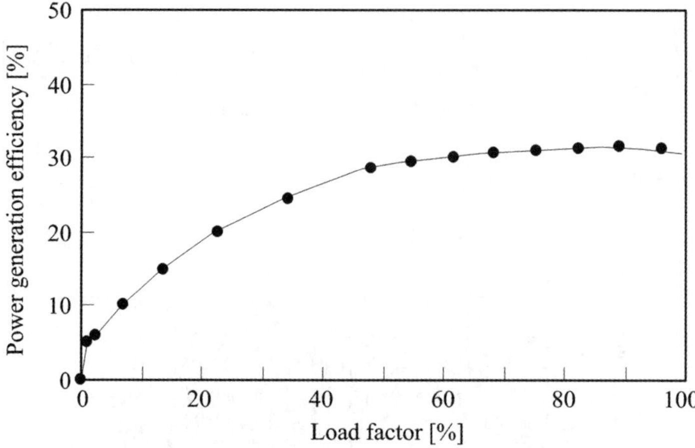

Figure 9.4 Output characteristics of a PEFC with city-gas reformer.

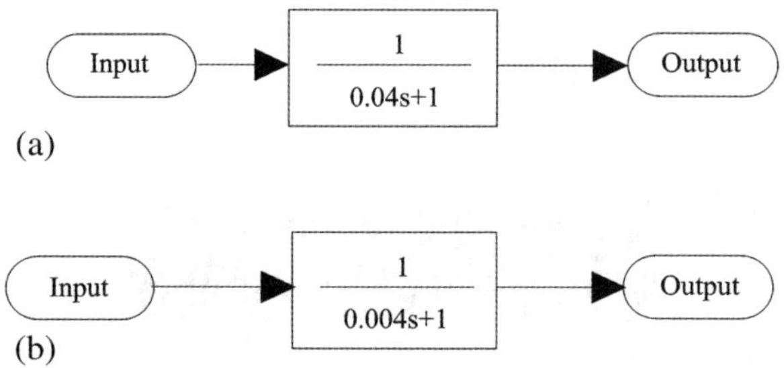

Figure 9.5 Transfer function of an inverter and interconnection device; (a) Transfer function of an inverter; (b) Transfer function of an interconnection device.

power output of the fuel cell system" by "the city gas calorific power supplied to the system". This model was prepared from the results of the power output when attaching the fuel cell shown in Figure 9.2(a) to the city-gas reformer shown in Figure 9.2(b). If the load of a fuel cell is given power-generation efficiency is calculable (Figure 9.4). The maximum efficiency of a fuel cell system is 32%.

9.3.5 Inverter and System Interconnection Device

It is assumed that an inverter of a voltage control type is used, and 120 ms is required to output power on regular voltage and frequency (in this chapter, it is less than 95%).[11] Figure 9.5(a) expresses the transfer function of such an inverter with primary delay.

When changing power with a system-interconnection device, the change takes about 10 μs.[11] However, there is the operation of taking the synchronism of the frequency between systems, and the model of the system-interconnection device sets the change time to 12 ms. As a result, the transfer function of the system-interconnection device by primary delay is shown Figure 9.5(b).

9.4 Control Parameters and Analysis Method

The response characteristics of the 1-kW fuel cell system when inputting a 0.2-, 0.6-, and 1.0-kW load stepwise is shown in Figure 9.6. The response characteristics of a fuel cell system changes by the control parameters set up with the controller. As shown in Figure 9.6(c), in 1-kW step input, the rising time and settling time (time to converge on $\pm 5\%$ of the target output) are not based on control parameters. In a 0.2-kW step input, the rise time of "$P = 12.0$, $I = 1.0$" is short, and the settling time of "$P = 1.0$, $I = 1.0$" is short. In a 0.6-kW step input, "$P = 12.0$, $I = 1.0$", and "$P = 1.0$, $I = 1.0$" have almost the same settling time. Moreover, overshooting is large although the rise time of "$P = 12.0$, $I = 1.0$" is short. Considering the following load fluctuations, the control parameters of the fuel cell are analyzed by "$P = 12.0$, $I = 1.0$". The dynamic characteristics of a microgrid are analyzed using MATLAB (Ver.7.0) and Simulink (Ver.6.0) of Math Work Corporation. However, in analysis, the solver to be used is the positive Runge–Kutta system, and this determines the sampling time from calculation converged to less than 0.01% by error.

9.5 Load Response Characteristics of the Microgrid

9.5.1 Step Response

The response results when applying the stepwise input of 2, 4, 6 or 8 kW to the microgrid at intervals of 30 s are shown in Figure 9.7(a). The left-hand side in Figure 9.7(a) shows the result of not installing a wind power generator. The right-hand side of the figure shows the result of a installing wind power generator. The maximum power by an overshooting and settling time (time to converge on $\pm 5\%$ of the target output) are described on the left-hand side of Figure 9.7(a). Moreover, the maximum power due to overshoot is described in the right-hand side figure. The settling time when not installing a wind power generator has the longest period of step input of 6 kW and 8 kW for 3.9 s. If a wind power generator is connected to the microgrid, many fluctuations in the system response characteristics will occur in a short period. If the power produced by wind-power generation is supplied to the microgrid, the dynamic characteristics of power of the microgrid will be influenced. Figure 9.7(b) shows the analysis result of the response error corresponding to Figure 9.7(a). If a wind power generator is connected to the grid, the response error will become large as the load of the grid becomes small. It is expected that the power range of the fluctuation of the microgrid will increase as the output of wind-power generation grows. Therefore, when the load of a microgrid is small compared

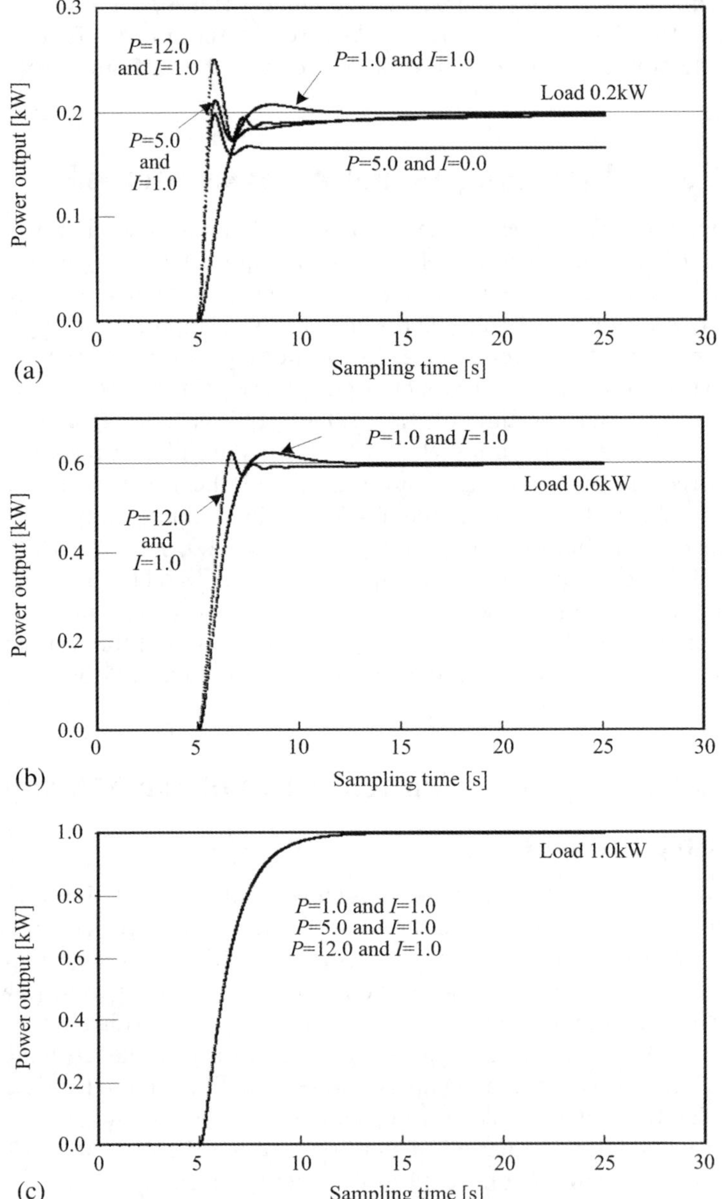

Figure 9.6 Characteristics of electric power output of the system; (a) Power load is 0.2 kW; (b) Power load is 0.6 kW; (c) Power load is 1.0 kW.

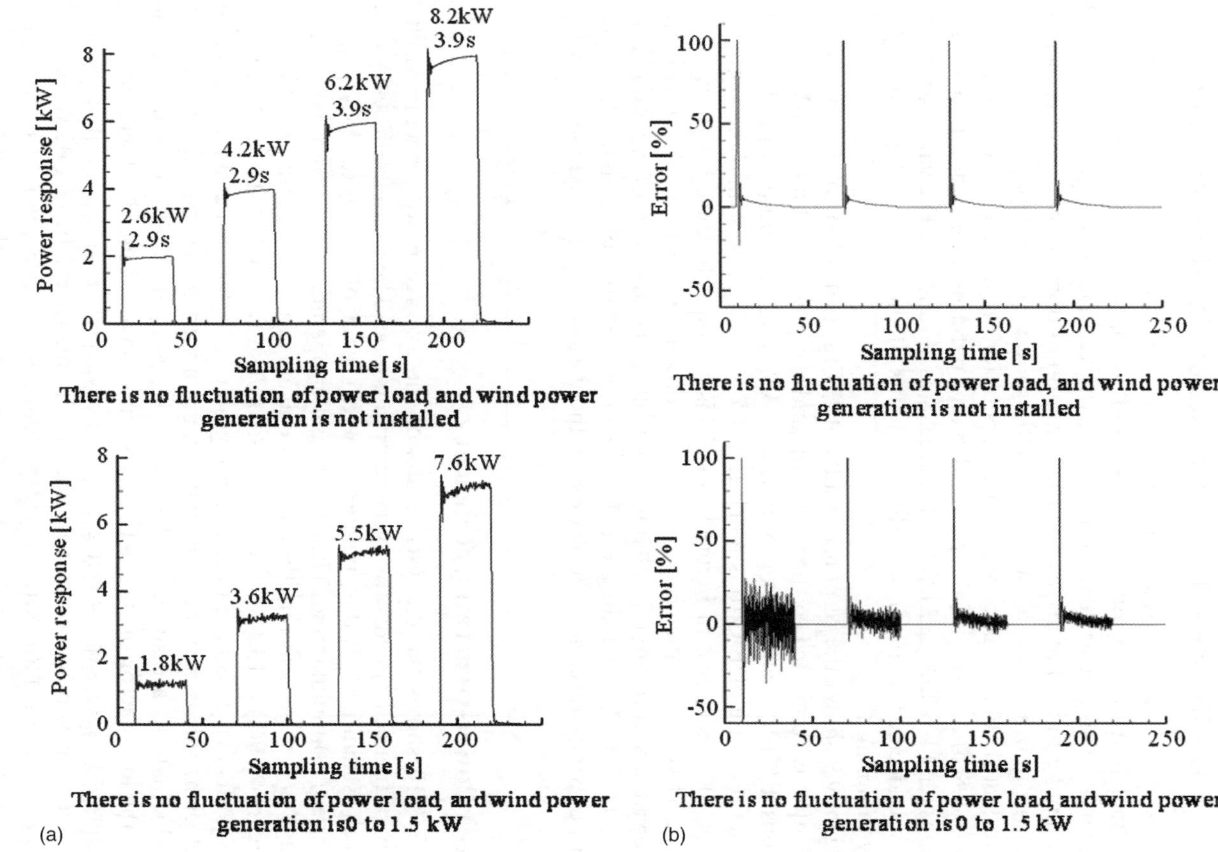

Figure 9.7 Results of step response; (a) Results of power response; (b) Results of input–output error.

with the output of a wind power generator, the power supply of the independent microgrid becomes unstable.

9.5.2 Load Response Characteristics of Cold-Region Houses

Figure 9.8(a) shows the power demand pattern of a microgrid formed from ten individual houses in Sapporo, Japan, and assumes a representative day in February.[12] This power demand pattern is the average value of each hour, and the sampling time of analyses and the assumption time are written together on the horizontal axis. As a base load of the power demand pattern shown in Figure 9.8(a), F/C(0) is considered as operation of a 2.5-kW constant load. Figures 9.8(b) and (c) are the power demand patterns when adding load fluctuations (±1 kW and ±3 kW) to Figure 9.8(a) at random. The variation of the load was decided at random within the limits of the range of fluctuation for every sampling time.

Figure 9.9 shows the response results of F/C(0) to F/C(6) when wind-power generation is connected to the microgrid and the power load has ±1-kW fluctuations. F/C(0) assumed operation with 2.5 kW constant output, with the result that the response of F/C(0) is much less than 2.5 kW in less than the sampling time of 100 s as shown in Figure 9.9(a). This reason is because F/C(0) was less than 2.5 kW with the power of wind-power generation. Although the microgrid assumed in this chapter controlled the number of operations of F/C(1) to F/C(7) depending on the magnitude of the load, since the power supply of wind-power generation existed, there was no operating time of F/C(7).

9.5.3 Power-Generation Efficiency

Figure 9.10 shows the analysis results of the average power-generation efficiency of fuel cell systems for every sampling time. The average efficiency of a fuel cell system is the value averaging the efficiency of F/C(0) to F/C(7) operated at each sampling time. The fuel cell under operation stop is not included in average power-generation efficiency. The average power-generation efficiency of Figure 9.10(a) is 13.4%, and Figure 9.10(b) shows 14.3%. The difference in average efficiency occurs in the operating point of a fuel cell system shifting to the efficient side, when load fluctuations are added to the microgrid. Thus, if load fluctuations are added to the microgrid, compared with no load fluctuations, the load factor of the fuel cell system shown in Figure 9.4 will increase.

Figure 9.11 shows the power-generation efficiency of each fuel cell in the case of connecting wind-power generation to the microgrid of ±1.0 kW of load fluctuation. F/C(0) operated corresponding to a base load has maximum power-generation efficiency at all sampling times. Since the number of operations of a fuel cell is controlled by the magnitude of the load added to the microgrid, the operating time falls in the order of F/C(1) to F/C(6). Moreover, there is no time to operate F/C(7) in this operating condition.

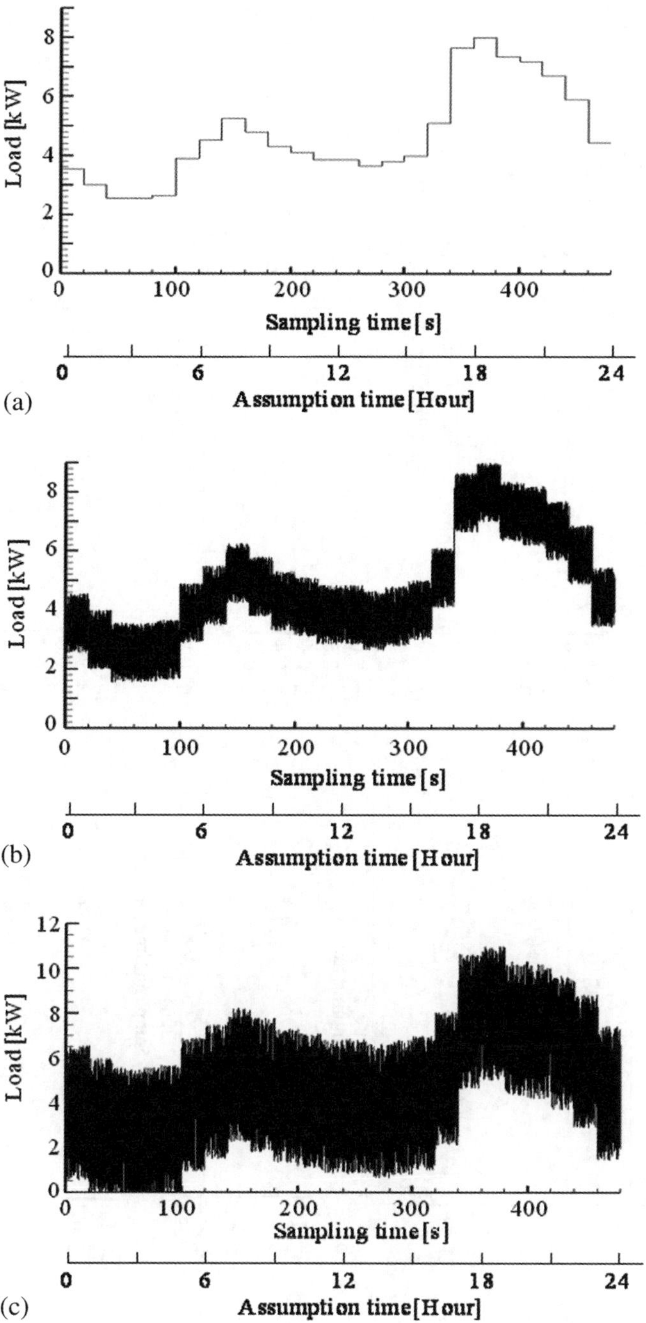

Figure 9.8 480 s demand model for 10 houses in February in Sapporo; (a) Average power load; (b) Fluctuation of power load is ±1.0 kW; (c) Fluctuation of power load is ±3.0 kW.

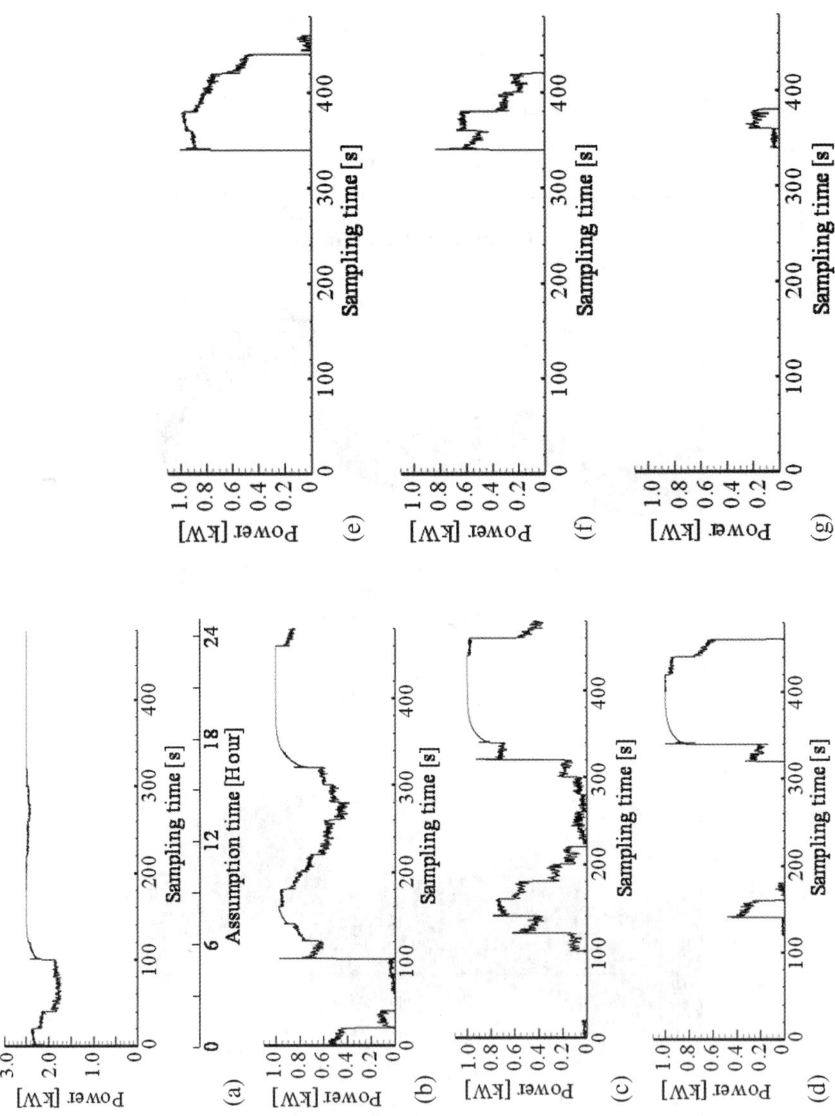

Figure 9.9 Response results of each fuel cell; (a) F/C(0); (b) F/C(1); (c) F/C(2); (d) F/C(3); (e) F/C(4); (f) F/C(5); (g) F/C(6).

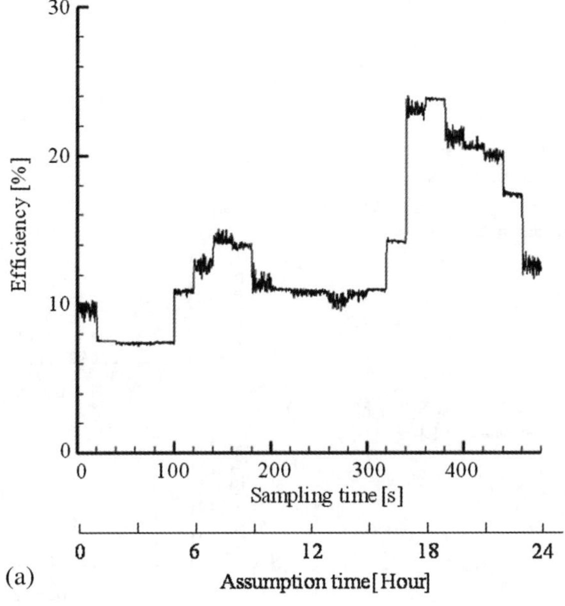

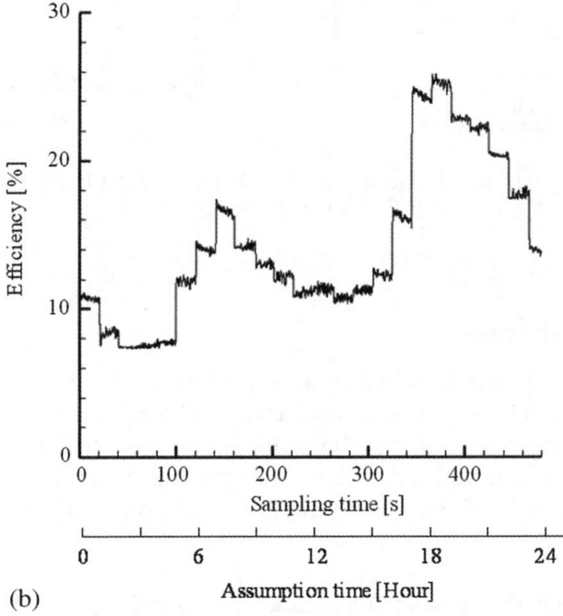

Figure 9.10 Results of microgrid average efficiency; (a) There is no fluctuation of power load and wind-power generation is 0 to 1.5 kW; (b) Fluctuation of power load is ± 3.0 kW and wind-power generation is 0 to 1.5 kW.

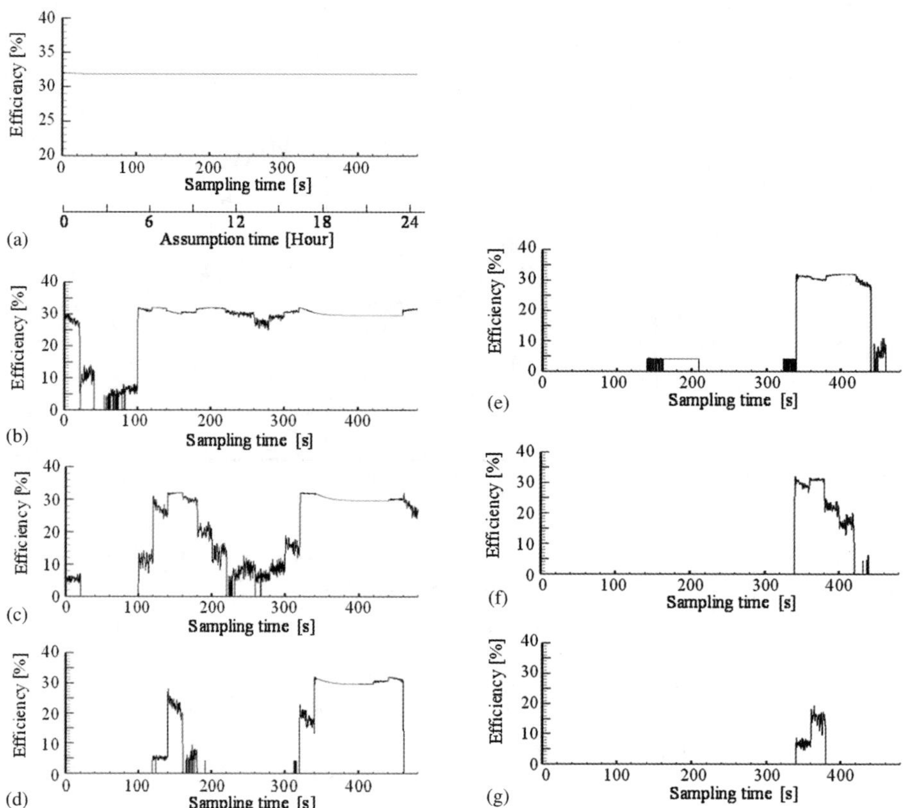

Figure 9.11 Results of efficiency for each fuel cell; (a) F/C(0); (b) F/C(1); (c) F/C(2); (d) F/C(3); (e) F/C(4); (f) F/C(5); (g) F/C(6).

9.6 Conclusions

A 2.5-kW fuel cell was installed in a house linked to a microgrid, operation corresponding to a base load was conducted, and the dynamic characteristics of the grid when installing a 1-kW fuel cell system in seven houses were investigated by numerical analysis. A wind power generator output to a microgrid at random within 1.5 kW was installed, and the following conclusions were obtained.

(1) Although the settling time (time to converge on ±5% of the target output) of the microgrid differs with the magnitude of the load, and the parameters of the controller, it is about 4 s.
(2) When connecting a wind power generator to the microgrid, the instability of the power of the grid due to supply-and-demand difference is an issue. This issue is remarkable when the load of an

independent microgrid is small compared to the production of electricity of unstable wind-power generation.

(3) When wind power equipment is connected to the microgrid with load fluctuation, the operating point of the fuel cell system may shift and power-generation efficiency may improve.

Acknowledgments

This work was partially supported by a Grant-in-Aid for Scientific Research(C) from the JSPS.KAKENHI (17510078).

Nomenclature

Act:	"if" action
Act_FC:	each fuel cell operation
F/C:	fuel cell
h:	capacity of generation W
I:	integral parameter
P:	proportionality parameter
PI:	proportion integration control
u:	power load of a microgrid W
v:	power output W

References

1. S. Abu-Sharkh, R. J. Arnold, J. Kohler, R. Li, T. Markvart, J. N. Ross, K. Steemers, P. Wilson and R. Yao, Can microgrids make a major contribution to UK energy supply? *Renew. Sustain. Energy Rev.*, 2006, **10**(2), 78–127.
2. A. Carlos and A. Hernandez, Fuel consumption minimization of a microgrid, *IEEE Trans. Indus. Appl.*, 2005, **41**(3), 673–681.
3. H. Robert, Microgrid: A conceptual solution, *Proceedings of the 35th Annual IEEE Power Electronics Specialists Conference*, 2004, **6**, 4285–4290.
4. S. Nagano, Plate-Type Methanol Steam Reformer Using New Catalytic Combustion for a Fuel Cell, *Proceedings of SAE Technical Paper Series, Automotive Eng.*, 2002, 10.
5. S. Obara and K. Kudo, Installation Planning of Small-Scale Fuel Cell Cogeneration in Consideration of Load Response Characteristics (Load Response Characteristics of Electric Power Output), *Trans. Jpn. Soc. Mech. Eng., Series B*, 2005, **71**(706), 1678–1685 (in Japanese).
6. B. Lindstrom and L. Petterson, Development of a methanol fuelled reformer for fuel cell applications, *J. Power Sources*, 2003, **118**, 71–78.

7. K. Oda, S. Sakamoto, M. Ueda, A. Fuji and T. Ouki, A Small-Scale Reformer for Fuel Cell Application, *Sanyo Technical Review*, 1999, **31**(2), 99–106, Sanyo Electric Co., Ltd., Tokyo, Japan (in Japanese).
8. Y. Takeda, Y. Iwasaki, N. Imada and T. Miyata, Development of Fuel Processor for Rapid Start-up, *Proc. 20th Energy System Economic and Environment Conference*, 2004, Tokyo, ed., K. Kimura, 343–344 (in Japanese).
9. S. Ibe, N. Shinke, S. Takami, Y. Yasuda, H. Asatsu and M. Echigo, Development of Fuel Processor for Residential Fuel Cell Cogeneration System, *Proc. 21th Annual Meeting of Japan Society of Energy and Resources*, 2002, Osaka, ed., K. Abe, 493–496 (in Japanese).
10. S. Obara and K. Kudo, Study on Small-Scale Fuel Cell Cogeneration System with Methanol Steam Reforming Considering Partial Load and Load Fluctuation, *Trans. ASME, J. Energy Res. Technol.*, 2005, **127**, 265–271.
11. Kyoto Denkiki Co., Ltd., A system connection inverter catalog and an examination data sheet; 2001.
12. K. Narita, *The Research on Unused Energy of the Cold Region City and Utilization for the District Heat and Cooling*, Ph.D. thesis, 1996, Hokkaido University (in Japanese).

CHAPTER 10

Solar Cell/Diesel Engine Compound System with Production-of-Electricity Prediction

SHIN'YA OBARA

10.1 Introduction

Microgrid technology with the capacity for sustainable energy operation has been widely discussed recently from the point of view of reducing the environmental impact of society.[1–3] In these setups, the operation optimization program installed in the controller of a combined system is the most important aspect of the technology for determining the performance of the system.[4] However, because an output prediction for the green energy contribution to the system is required, the dynamic operation plan of a system that combines conventional energy equipment (for example, a diesel engine, a gas engine, a fuel cell, *etc.*) and green-energy equipment can be very difficult to design. In this work, we use a neural network (NN) to obtain output predictions for a solar cell. Weather data from the past 14 years (amount of solar radiation and outside temperature) is fed into the learning process of the NN. This NN production-of-electricity prediction algorithm was developed by the author and is described as PAS in ref. 5.

Power fluctuations are known to occur in systems that utilize green energy on an independent microgrid and that experience large or rapid changes in load.[6] Given this, power storage equipment must be introduced and the dynamic characteristics of the microgrid must be improved. Due largely to the

RSC Energy Series No. 3
Compound Energy Systems: Optimal Operation Methods
By Shin'ya Obara and Arif Hepbasli

Published by the Royal Society of Chemistry, www.rsc.org

proliferation of hybrid vehicles and the like, the cost and performance of batteries have recently improved remarkably.[7] With this in mind, this chapter investigates algorithms for the operation planning of a microgrid that combines conventional type energy equipment, a solar cell and a battery. Since a microgrid is typically built up of two or more energy systems, we have to solve a nonlinear problem with many variables. In this chapter, the operation condition of the generating equipment is expressed with chromosome codes, these are introduced into a GA, and the optimal operation planning is determined.

10.2 Independent Microgrid with Renewable Energy and Battery

10.2.1 System Configuration

Figure 10.1 shows the case of an independent microgrid that distributes power to a small collection of generating equipment (here we use gas engine generators as an example, but our analysis is not limited to this case). The microgrid is controlled by a system controller on which we have installed dynamic operation planning software. In this chapter, we investigate the power supply-and-demand characteristics of the microgrid; the examination concerning the supply-and-demand of heat in a similar system will be reported independently.

10.2.2 Dynamic Operation Planning

(1) Operation planning based on a solar cell output power model.
Figure 10.2 shows the operation method of a microgrid with a solar cell. Dynamic operation of the microgrid is planned based on the output power

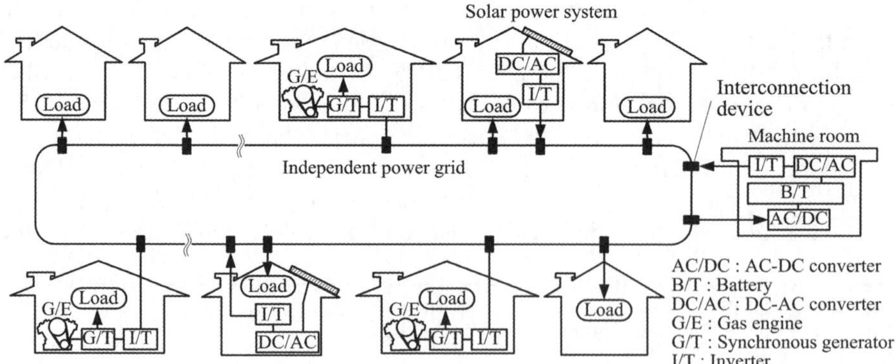

Figure 10.1 Independent microgrid system using gas engines.

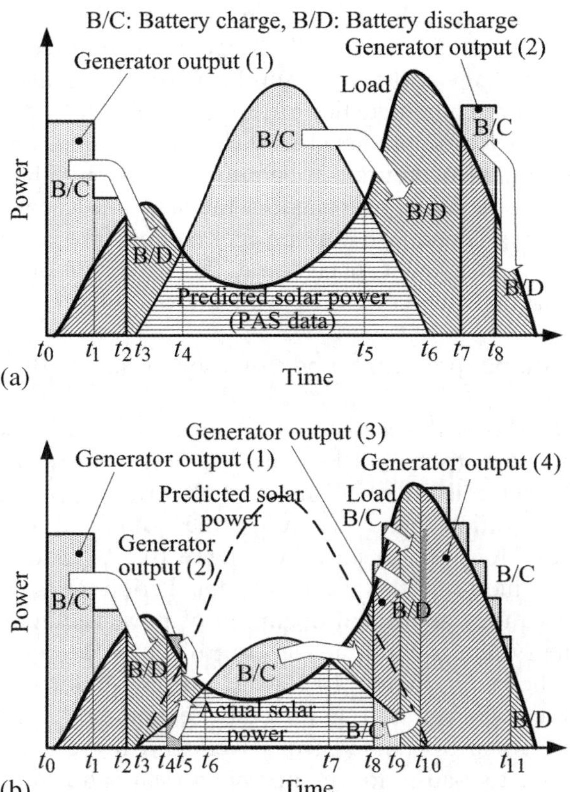

Figure 10.2 Operation planning of a microgrid with a solar power; (a) Operation planning based on the PAS predictive data; (b) Operation planning when taking into consideration the error between PAS predictive data and actual solar power.

model of a solar cell. In this chapter, the solar cell output model is based on PAS predictive data and past average weather data.[8,9] The PAS is a production-of-electricity prediction algorithm that was designed specifically for the solar cell in the work.[5] Figure 10.2(a) shows the method of planned dynamic operation based on the output power model of the solar cell described in the introduction. From time t_0 to time t_2, there is no output from the solar cell and so in this interval, more electric energy is stored in the battery than is required by the load. In light of this, generating equipment is operated during this period (Generator output (1)). Each piece of generating equipment is operated by the fixed load near the maximum efficiency point. The supply of electric power in the microgrid is adjusted by controlling the number of pieces of power equipment in operation. When the supply of electric power exceeds the load in the period from t_0 to t_2, the amount of surplus charge is moved into the battery. There is an output of the solar cell in the period t_3 to t_6. In the period t_2 to t_4, the system responds to a load by discharging the battery, and supplying charge from output of the solar cell. Because the output of the solar cell exceeds the load in the period t_5 to t_4, surplus power can again be stored in the battery. If the output characteristics of the solar cell and the characteristic of power load are able to be predicted in the early morning, one can minimize the number of equipment hours of operation and the number of charge and discharge cycles on the battery throughout the day. It is known, however, that the load following operation of power equipment is disadvantageous when it comes to installed capacity and the hours of operation required of power equipment when compared with the output predictive model of a solar cell.[5]

(2) Error of the output predictive model of a solar cell, and the influence on an operation plan.

Figure 10.2(b) shows the method of an operation plan when a difference occurs between the output model of the solar cell shown in Figure 10.2(a), and an actual solar cell output. This system starts out (during t_0 to t_2) by following the operation plan shown by the command of the controller in Figure 10.2(a). However, as shown in Figure 10.2(b), since there is so little power being produced in the solar cell, the period of t_4 to t_5 shown in Figure 10.2(b) requires additional operation of power equipment (Generator output (2)). The system is continuously controlled by the controller to follow the operation shown in Figure 10.2(a). However, there is little storage of electricity by the solar cell in Figure 10.2(b) compared with Figure 10.2(a). For this reason, when discharged from a battery in the period t_7 to t_8, the system will change to load-following operation (generator output (3) and (4)). As a result, the number of hours of operation of power equipment increases, and the additional capacity of the battery cannot be sufficiently utilized further.

10.2.3 Solar Cell System

The area of power generation of the solar cell introduced into the microgrid will be called S_{sol} and we are assuming a polycrystalline-silicon-type solar cell. The production of electricity of the solar cell P_{sol} is calculated using eqn (10.1). The power-generation efficiency changes as the temperature of the solar cell T_{ref} changes (the efficiency falls as the temperature increases). The temperature coefficient in this case is called R_T. T_∞ is a reference temperature and η_{sol} is the power-generation efficiency of the solar cell at T_∞. H_D in eqn (10.1) expresses the solar radiation intensity for direct delivery (the intensity of radiation which enters into the acceptance surface). Moreover, H_S in eqn (10.1) expresses the solar radiation intensity of the dispersion component,

$$P_{sol} = S_{sol} \cdot \eta_{sol} \cdot (H_D + H_S) \cdot \{1 - (T_{ref} - T_\infty) \cdot (R_T/100)\} \tag{10.1}$$

10.3 Power Balance and Objective Function

10.3.1 Power Balance

Equation (10.2) expresses the power-balance equation in the proposed microgrid. The left-hand side of the equation is the power output by the composition of equipment that makes up the system, and the right-hand side expresses the power consumed by the microgrid. $E_{gen,i,t}$, $E_{bt,t}$ and $E_{sol,t}$ on the left-hand side express the output of the generating equipment, battery, and solar cell between time t and $t+1$, respectively. Moreover, N_{eng} is the number of pieces of generating equipment introduced into the microgrid. $E_{need,j,t}$ is the power demand at time t of the house j. N_{house} is the number of the houses connected to the microgrid. The last term on the right-hand side of eqn (10.2) ($\Delta E_{loss,t}$) expresses the power loss in the system. The charge-and-discharge efficiency of a battery, power transmission loss, *etc.*, are included in this term. In the analysis of this chapter, the charge-and-discharge efficiency of a battery is the only effect included in this term. The sampling interval for each piece of equipment, such as the generating equipment, a solar cell, or a battery, is set to be one hour in this chapter.

$$\sum_{i=1}^{N_{eng}} E_{gen,i,t} + E_{bt,t} + E_{sol,t} = \sum_{j=1}^{N_{house}} E_{need,j,t} + \Delta E_{loss,t} \tag{10.2}$$

10.3.2 Objective Function

The number of hours of operation of the generating equipment EOT_{t} between time t to $t+1$ is obtained by calculating eqn (10.3). However, $N_{gen,i,t}$ expresses the operational status of the generating equipment i in the time between t and $t+1$ ("1" indicates operation, "0" indicates idle). N_{gen} is the number of pieces of generating equipment installed in the microgrid. The total number of hours of operation for all pieces of generating equipment that were in operation during the periods $t = 1,2, \ldots, P_{sys}$ is calculated by eqn (10.4). The optimization of the dynamic operation plan of the microgrid is examined using a genetic algorithm (GA). In the GA, the objective function shown by eqn (10.4) is defined as an adaptive value. The solution that more closely satisfies eqn (10.4) is described as having a "large adaptive value."

$$EOT_t = \sum_{i=1}^{N_{gen}} N_{gen,i,t} \tag{10.3}$$

$$\text{Engine operation time} = \text{minimize}\left(\sum_{t=1}^{P_{sys}} EOT_t\right) \tag{10.4}$$

10.4 Analysis Method

10.4.1 Production-of-Electricity Prediction Algorithm of Solar Cell (PAS)[5]

A layered neural network (NN) is introduced and the production of electricity of a solar cell is predicted according to the following procedures.

(1) Input-and-output data of NN used by PAS.

Figure 10.3 expresses the input-and-output data of the NN used for PAS. *dw* expresses the present date and *t* expresses the present time. In the learning and analysis process of the NN, the average amount of solar radiation and average outdoor air temperature are input as data for each time of the present date. The input data described in the introduction is fed into eqn (10.1), and the production of electricity P_{sol} of the solar cell is obtained. This P_{sol} is used as the teaching data in the learning process of the NN.

(2) Input data.

The data input into the NN the learning process includes the average amount of solar radiation and average outdoor air temperature for each time of the day (from time 0 to time *t*). On the other hand, the average amount of solar radiation and average outdoor air temperature from *t* to 24 of a given day give the values measured on the same time and the same day of previous years, as obtained from "the standard weather and the solar radiation database on weather government office and AMEDAS (1990 to 2003)",[8] and "NEDO technical information data base (METPV-3)".[9] The data input into the NN during the analysis process are the same as those input during the learning process from 0 to *t*. The average amount of solar radiation and average outdoor air temperature from *t* to 24 of a given day give the data measured at same time the previous day.

(3) Structure of the NN, and output data.

Figure 10.4 shows the structure of the NN introduced into PAS. Figure 10.4(a) shows the learning process and Figure 10.4(b) shows the analysis process. The NN used in this proposal has three layers. In the learning process, the input data (x_1 to x_{48}) (described in Section (2)) are fed into the first layer (the power-input layer) and the teaching data (y_1 to y_{24}) are fed into the third

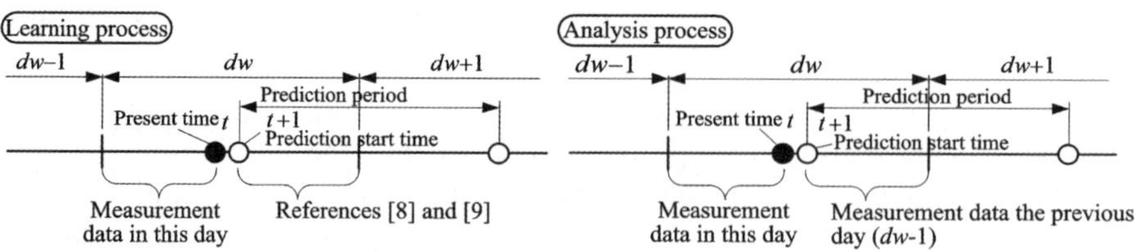

Figure 10.3 Input data introduced into proposal NN.

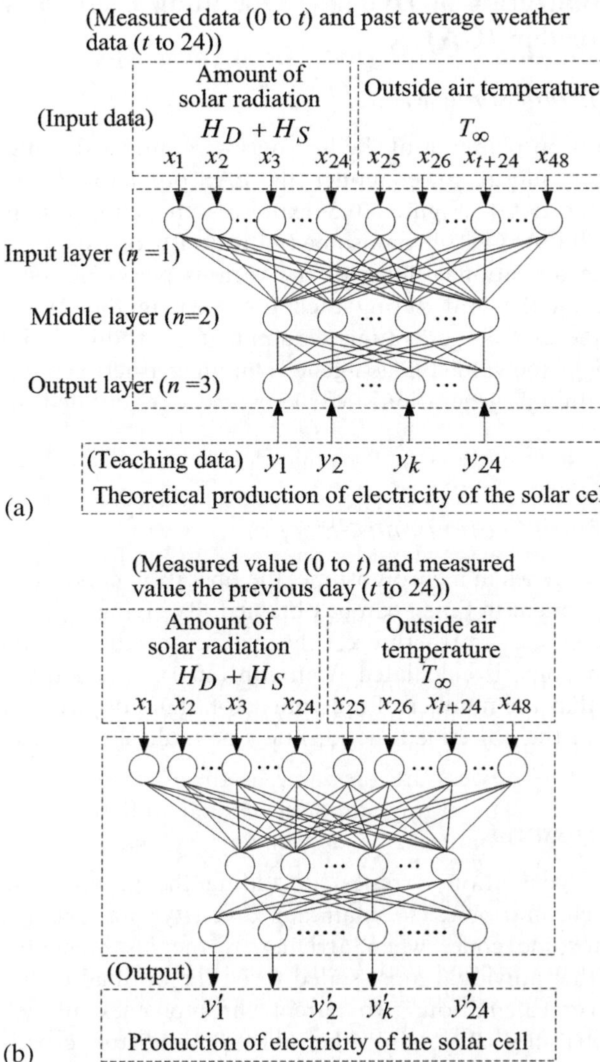

Figure 10.4 Layered neural network of the proposed system; (a) Learning process; (b) Analysis process.

layer (the output layer). The weight of each network connection between neurons is decided using backpropagation[10] so that the relationship between each input data point and each teaching data point may be realized. Input data can be given to a learned NN (in the analysis process), and the solar cell output power (y'_1 to y'_{24}) in each time of *dw* can be obtained from the output layer.

10.4.2 Optimization of Dynamic Operation Using a Genetic Algorithm (GA)

10.4.2.1 Chromosome Model

Dynamic operation planning of the microgrid is optimized using a GA based on the solar cell output power model (the PAS predictive data and the past average weather data). Figure 10.5 expresses the chromosome code used by the proposed GA. One chromosome (individual) consists of 24 genes. Each gene shows the operational status of the various pieces of generating equipment that are on the grid during each time step in the date *dw*. With the number of pieces of generating equipment introduced into the grid being called N_{gen}, the value of each gene is an integer between 0 and N_{gen}. The genes of the initial generation's chromosome are decided using random numbers.

10.4.2.2 Multiplication and Selection

By decoding the genes in a chromosome, the operation condition $N_{gen,i,t}$ of the generating equipment in time t can be obtained. By giving $N_{gen,i,t}$ to eqn (10.3), EOT_t is calculable. Furthermore, the value of the objective function of each chromosome is calculated from eqn (10.4). The software implementation is such that the number of individuals of high adaptive value (obtained by evaluating of the objective function) may be multiplied at a fixed rate.

10.4.2.3 Crossover

The change of generation is repeated, adding the genetic manipulation of crossover to chromosomes to maintain diversity (an evaluative process). In the chromosome code, when arriving at the last generation, a fitness value decides the individual most suited to be the optimal operating method. In the crossover calculation, two parent chromosomes are selected by the crossover probability P_{cros} given beforehand, and the crossover position common to both parent chromosomes is decided at random. The genes of both of the parent chromosomes are rearranged bordering on the crossover position, and the chromosome with a new gene is generated.

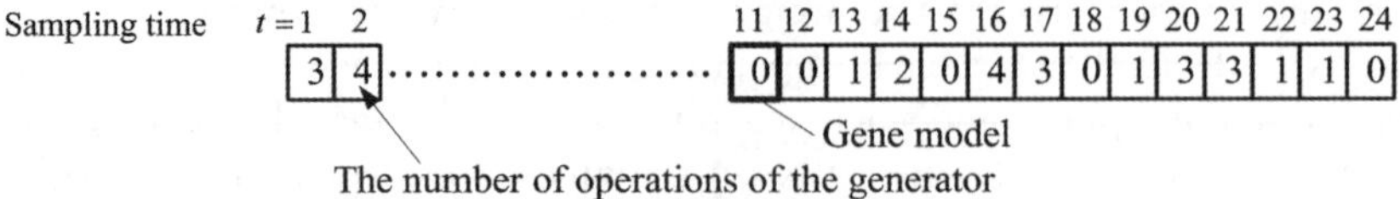

Figure 10.5 Chromosome code used for genetic algorithm.

10.4.3 Analysis Flow of Operation Planning

Analysis flow of the operation planned optimization of the microgrid by the proposal GA is shown in Figure 10.6. In the calculation part (a) of the figure, an electricity demand pattern, a solar cell output power model (PAS or output power pattern of the solar cell based on the past average weather data), and the parameter of GA are input. The initial generation's chromosome group is generated in calculation part (b). In calculation part (c), the fitness value of all the chromosomes is calculated and the order of chromosomes is decided at calculation part (d) according to the magnitude of their fitness values. Chromosomes of large fitness value are made to increase in number at a fixed rate, while chromosomes of small fitness value are deleted. In calculation part (e), a parent chromosome is chosen at random under crossover probability, a crossover position is decided at random, and genes are exchanged. In calculation parts (c), (d), (e), the calculation of the fitness value, the operation of arranging chromosomes in order of fitness value, and exchange of the gene by

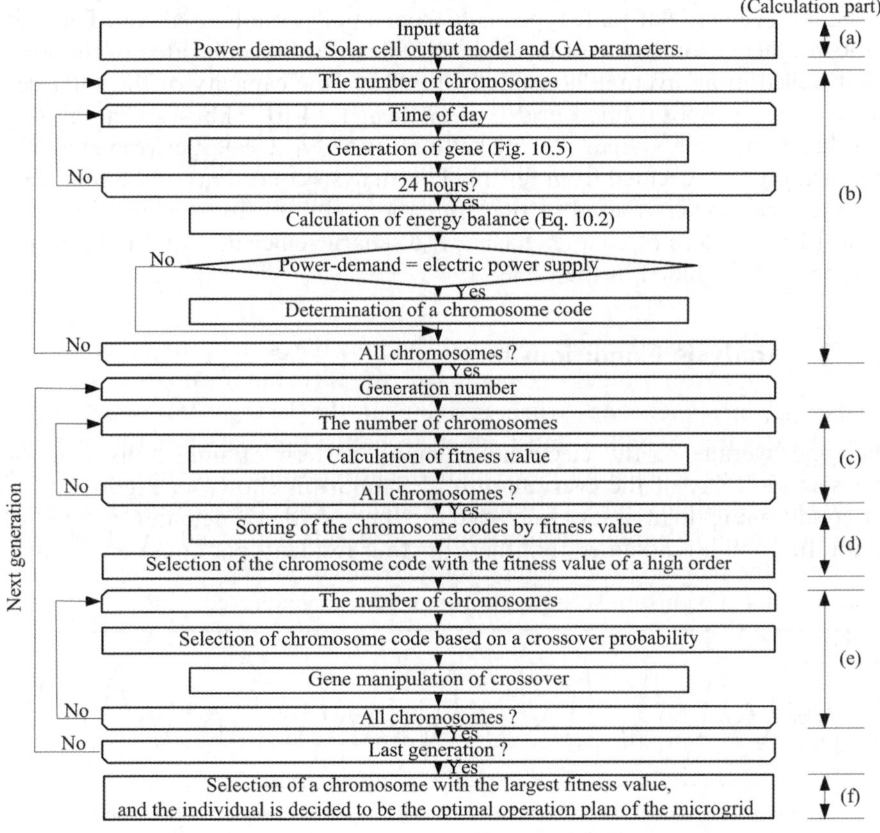

Figure 10.6 Calculation flow of the optimal operation plan using GA.

crossover is done. These operations are calculated and repeated until we reach the final generation number. In the last generation's chromosomes, an individual with the highest fitness value is decided to be the optimal solution. By decoding the chromosome code of the optimal solution, operation of the generating equipment in each time step of date *dw* can be planned.

10.5 Case Analysis

10.5.1 Analysis System

In this case analysis, we assume the introduction of the microgrid into nine houses in Sapporo. In this section, the difference of the operation plan when introducing the proposal method or the past average weather data is investigated. The generating equipment introduced into the microgrid will be structured in five sets, and the power output of each piece of generating equipment will be 1 kW. As described in Section 10.5.2, the power load introduced into this case analysis has a maximum of about 7.5 kW (Figure 10.7).

The solar cell is a flat-plate type made from a polycrystalline silicon. The area is 150 m^2, facing south, the slope angle is 30 degrees. Since the area of the solar cell installed in an average house is 20 to 40 m^2, the capacity of the solar cell linked to the proposal microgrid is equivalent to 4 to 7 houses. The battery introduced into the system is a nickel-hydrogen type. The performance and specification were decided from ref. 10. The analysis conditions of the solar cell, battery, each converter used in the simulation, and an inverter are shown in Table 10.1. The loss of charge efficiency, discharge efficiency, and natural discharge are all included in the table.

10.5.2 Analysis Conditions

(1) Power load.

The time average of the electricity demand pattern of nine houses in the representation day of the every month of Sapporo is shown in Figure 10.7.[11] Air-conditioning load is not included in the electricity demand pattern of Figure 10.7 and it is assumed that there are four residents per house on average.

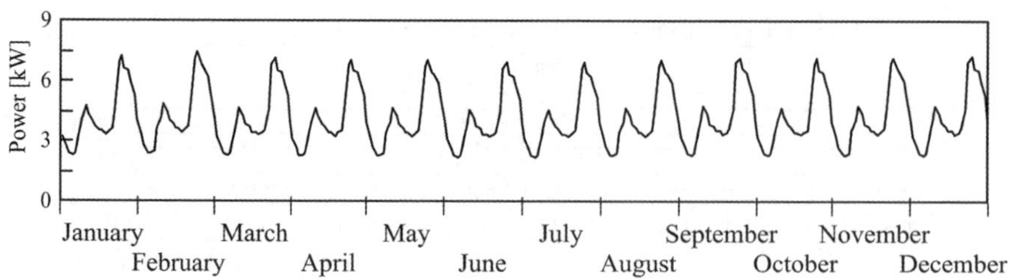

Figure 10.7 Power demand of the nine-house microgrid in Sapporo.

Table 10.1 Specifications of equipment.

Solar cell type	Multicrystalline silicon
Generation efficiency of the solar cell	14%
Temperature coefficient of the solar cell	0.4%/K
Battery type	Nickel-hydrogen
Battery efficiency	90%
Efficiency of AC–DC converter	95%
Efficiency of DC–AC converter	95%
Efficiency of DC–DC converter	95%

Space heating load is supplied with engine exhaust heat, and cooling load is unnecessary. Although the actual electricity demand pattern changes sharply on short time scales, a gradually varying time-averaged value is used throughout this chapter.

(2) Amount of solar radiation and outdoor air temperature.

In this case analysis we will investigate about six days, from the 2nd day to the 7th day in each of March, June, September, and December (the 1st day of every month is used to check the prediction of the PAS analysis). Figures 10.8 and 10.9 show the measured amount of solar radiation and outside temperature for seven days in every month from 1990 to 2004 in Sapporo.[8,9] The daily fluctuations in the amount of solar radiation are typically large when compared to the daily fluctuations of the outdoor air temperature in every month.

(3) GA parameters.

In this case analysis, the number of an initial generation's chromosome codes is 8000. In the genetic manipulation of multiplication and selection, the 20 chromosome codes (that is, 5% of the population) with the largest fitness values are made to increase their number in the next generation. The crossover probability is set to 0.2. The generation number was set to 10 because of the large total number of chromosome codes. These parameters of GA were arrived at by a trial-and-error method, and the final values fixed using analysis accuracy as a reference.

10.6 Analysis Results

10.6.1 Prediction of Solar Cell Output Power via PAS

(1) Relationship between prediction start time and analysis accuracy.

Figure 10.10 shows the predicted amount of solar radiation on June 6, 2007 in a south-facing set of 30 distinct angular orientations, calculated using PAS. In Figure 10.10, it will be the present time in 5, 8, and 11, and will be each prediction start time in 6, 9, and 12. Actual data refers to the weather data measured on June 6, 2007. We see that the difference between past average weather data[8,9] and actual data is larger than the difference between the actual data and

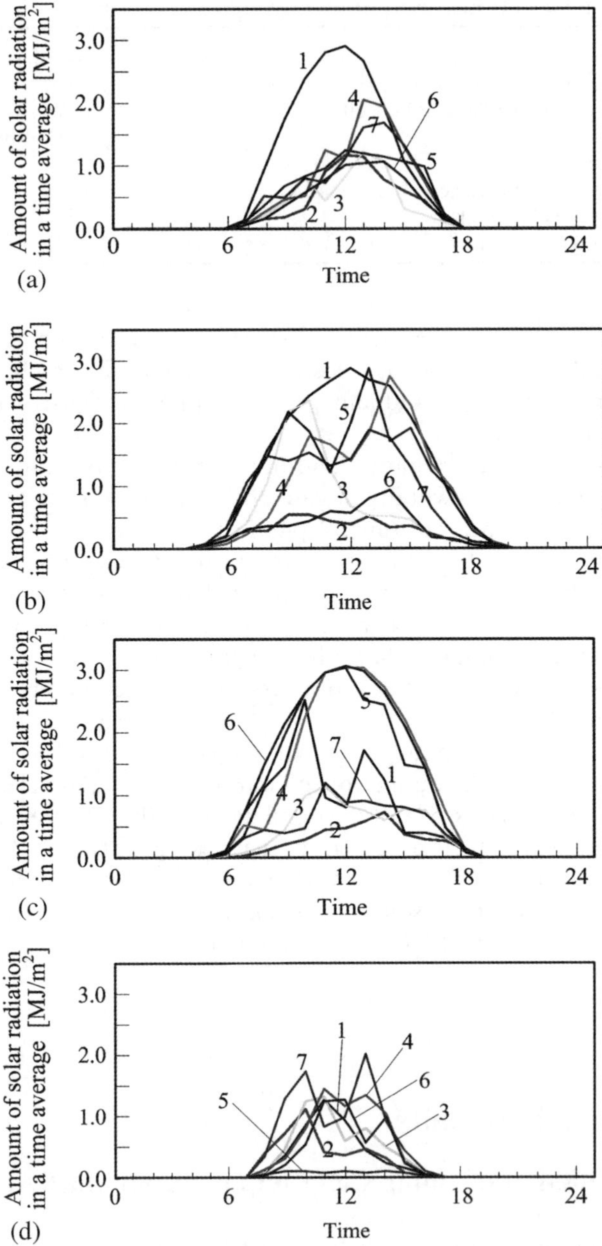

Figure 10.8 The slope-face amount of solar radiation in Sapporo in 1990 to 2004 (south-facing, the slope angle is 30 degrees; (a) March; (b) July; (c) September; (d) December.

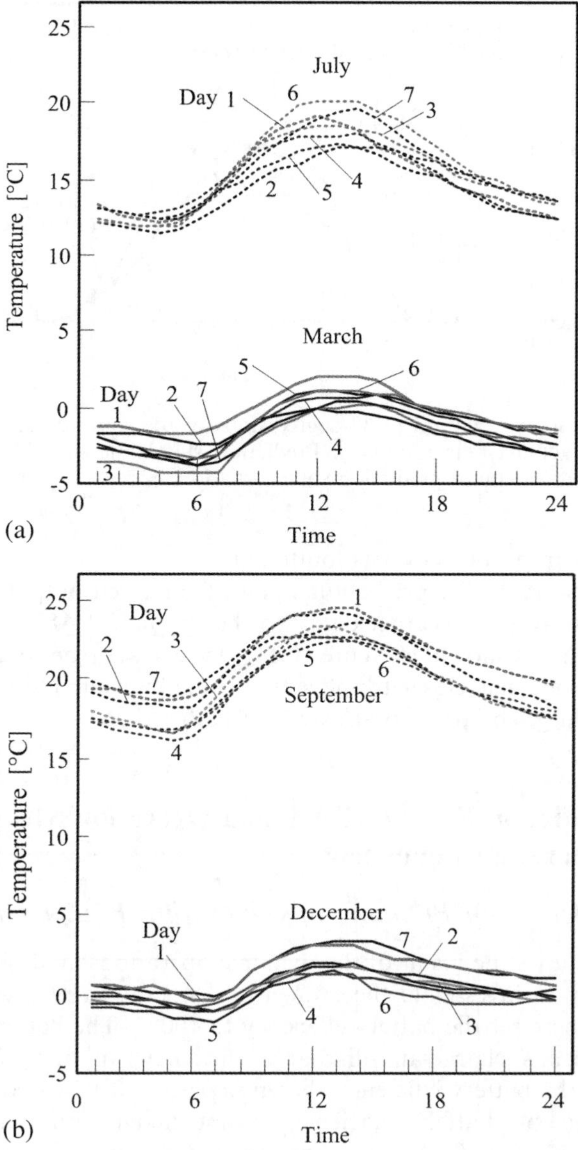

Figure 10.9 Outside temperature data in Sapporo in 1990 to 2004; (a) March and July; (b) September and December.

the predicted results. Moreover, the data that came from using a start time of 9 or 12 are closer to the actual data than that which came from using a start time of 6. In the result of Figure 10.10, if there is a large power input when using the weather data measured at that day (accordingly, the prediction start time is late), the analysis accuracy will have improved. In the analysis using PAS or later, prediction start time is 6 o'clock.

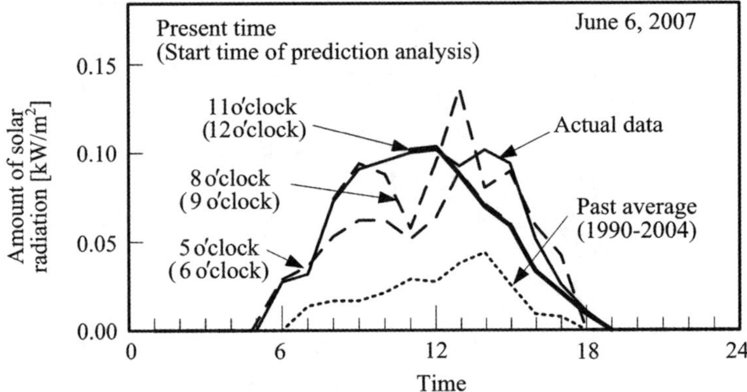

Figure 10.10 Solar radiation in 30-degree slope of the south sense using PAS, and actual solar radiation. Prediction start time of PAS, and the characteristic of prediction solar radiation.

(2) Prediction result in every month.

The result based on the prediction result of solar cell output power, actual value, and past average weather data[8,9] of having used PAS in representation days every month is shown in Figure 10.11. One can see from Figure 10.11 that in many cases one gets closer to the actual value by using the PAS prediction than one would get by using past average weather data.

10.6.2 Prediction Error of PAS, and Operation Method of Generating Equipment

10.6.2.1 Operation Planning of Generating Equipment

Figure 10.12 shows the result of the dynamic operations analysis on March 2, 2007. In order to investigate the influence of operation planning on battery efficiency, we here set the battery efficiency to 100%. The battery efficiency in Table 10.1 means charge-and-discharge efficiency. In analyses other than Figure 10.12, the battery efficiency shown in Table 10.1 is used. Accordingly, the loss based on battery efficiency is not taken into consideration in Figure 10.12. Figure 10.12 shows the calculation result except charge-and-discharge loss of the battery, in order to clarify the power balance about consumption, the solar cell, and the generator equipment. Figure 10.12(a) shows the result of operation planning of generating equipment using the solar cell output power prediction generated by PAS. On the other hand, Figure 10.12(b) shows the result of operation planning of generating equipment using the actual amount of solar radiation and outdoor air temperature. If the PAS-generated solar cell output power is the same as that coming from actual data then the optimal operation planning method is shown in Figure 10.12(a). However, one must also note that errors are introduced

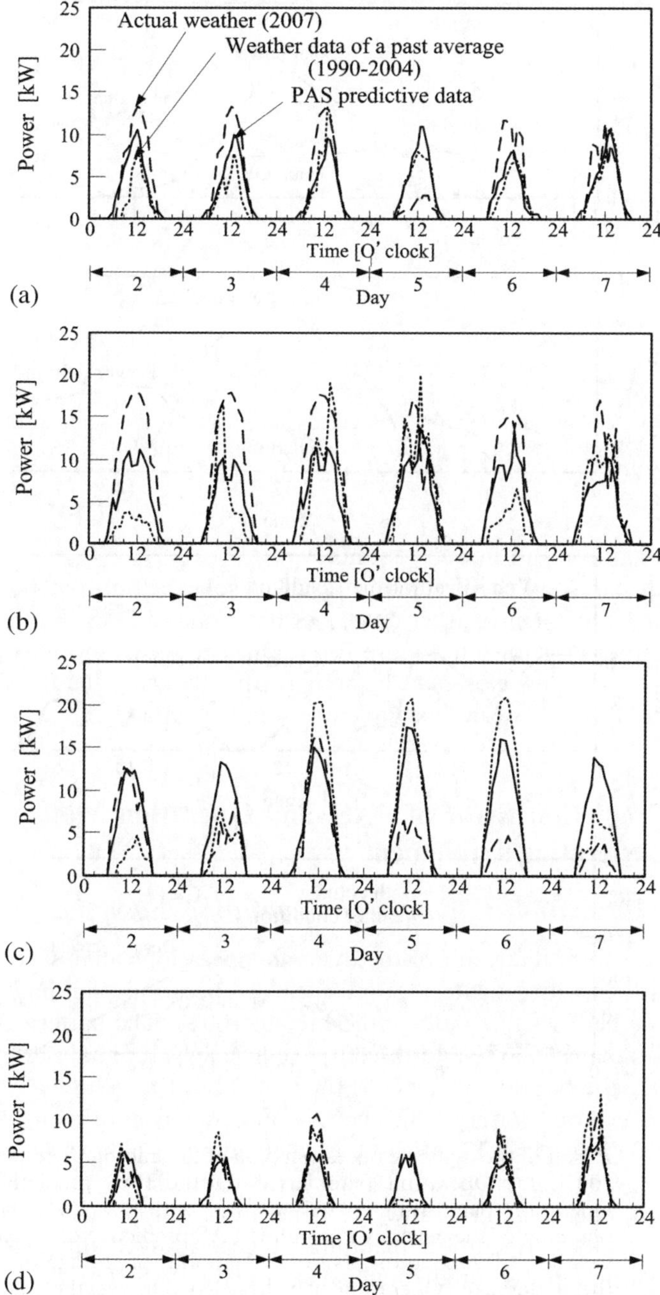

Figure 10.11 PAS prediction results and the actual value of the solar cell output in Sapporo; (a) March; (b) June; (c) September; (d) December.

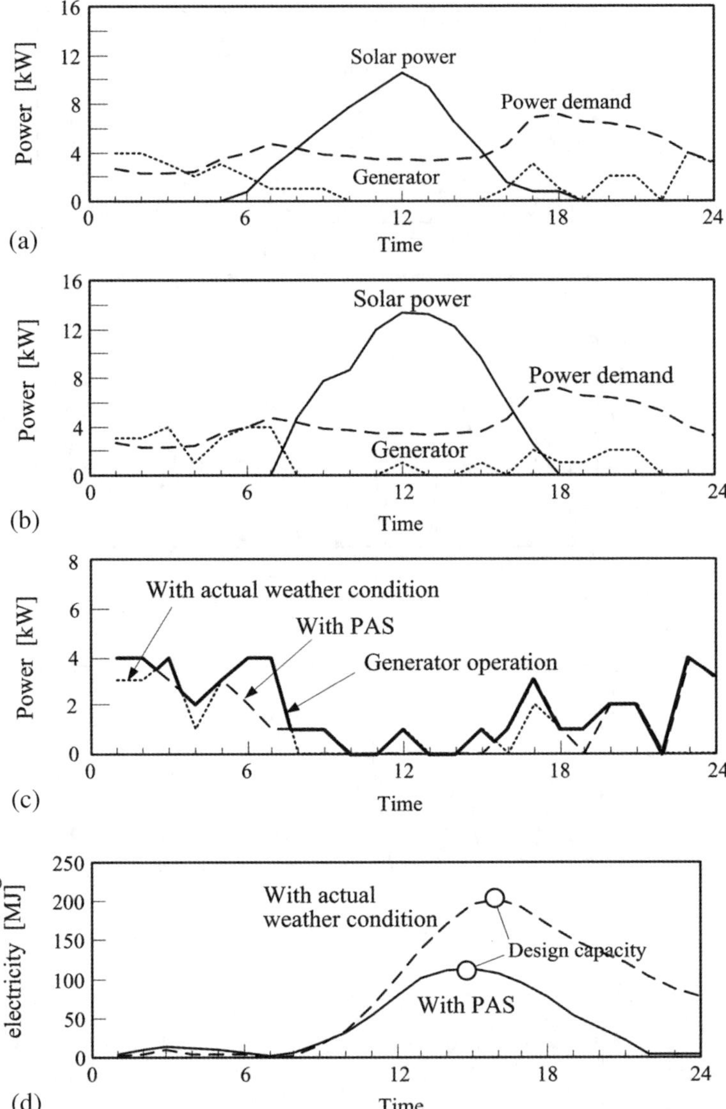

Figure 10.12 Result of the dynamic operation plan of the microgrid system (March 2, 2007); (a) Operation plan based on the PAS predictive data; (b) Operation plan using actual weather data; (c) Dynamic operation planning of the generator using the PAS predictive data and the actual weather; (d) Operation plan of the amount of storage of electricity in the battery; (e) Operation plan based on the weather data of a past average; (f) Operation plan of the amount of storage of electricity in the battery based on the weather data of a past average; (g) Dynamic operation planning of the generator using the weather data of a past average and the actual weather.

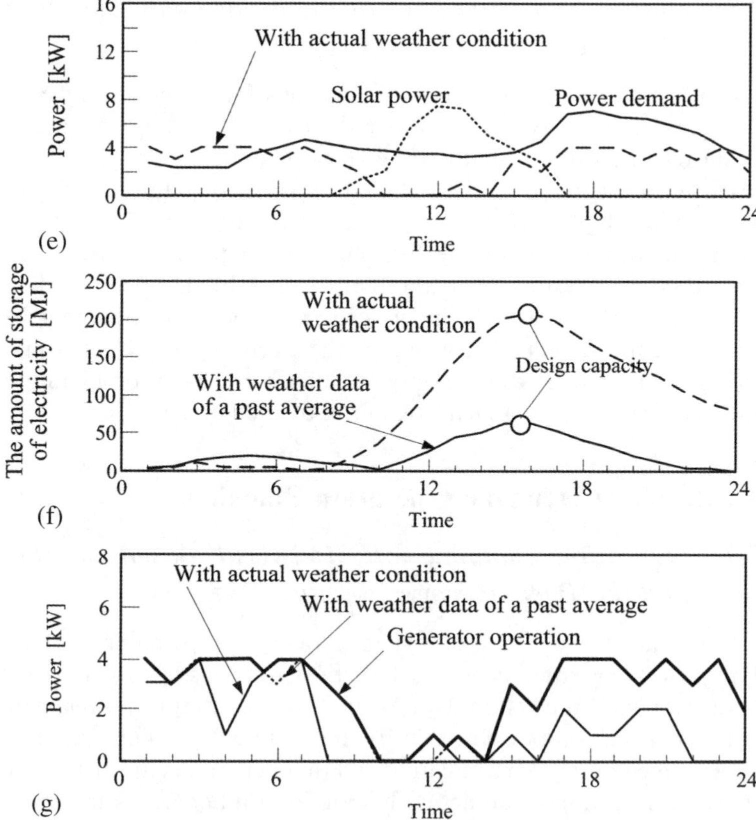

Figure 10.12 Continued

in the PAS predictions and actual operation may more closely resemble the result of Figure 10.12(c). Accordingly, the output power of the generating equipment as controlled by PAS and that controlled by actual weather conditions are compared, and the system is operated in such a way as to accommodate the worst conditions (this is a typical way of doing things in similar systems).

10.6.2.2 Operation Planning of Battery

Figure 10.12(d) shows the results that come from the operation planning of a battery when using either PAS predictions or actual data to calculate the solar cell output power. One sees that the operating characteristics of the battery greatly influence the method and duration of operation of the generating equipment. Looking at Figure 10.12(d), one sees that, in at least two different places, there is a significant difference in the battery capacity calculated in the two different schemes.

10.6.2.3 Output Characteristics of the Solar Cell, and Operation Planning

Figure 10.12(e) shows the operation planning of the generating equipment when using either past average weather data or actual weather data. Figure 10.12(f) shows the resultant battery capacity planning under the same conditions. Figure 10.12(g) shows the result of operation planning of generating equipment. When Figure 10.12(c) is compared with Figure 10.12(g), one sees that distinct methods of planning for generating equipment begin to deliver noticeably divergent results after about time 15. This happens as the output power models of the solar cell introduced into operation planning begin to produce diverging results. Accordingly, the accuracy of the output power prediction generated by PAS has a large influence on operation planning of the generating equipment the capacity planning of the battery.

10.6.3 Result of Dynamic Operation Planning

10.6.3.1 Operation Planning of Microgrid Based on the PAS and the Past Average Weather Data

Figure 10.13 shows the result of dynamic operation planning of the system (including generating equipment and batteries) when using either past average weather data or PAS prediction data as the solar cell output power model. The error in the solar cell output power in Figure 10.13 is the difference between the output determined by a given simulation and that determined from actual weather conditions. In operation of an actual microgrid, a difference is in the solar cell output power model described in the introduction, and the solar cell output power operating under actual weather conditions. As Section 10.6.2.1 described, in dynamic operations planning using the solar cell output power model, additional operation of generating equipment is expected in unfavorable conditions. For example we see, in Figure 10.13(b), that the partial output power of the generating equipment on December 5 (calculated in the past average weather data scheme) exceeds 5 kW, the maximum power of the proposed system. We can make sense of this as the smallest amount of solar radiation comes in the month of December. Because of this, the solar cell output power reduces and the discrepancy with past average weather data increases. Thus, if the error between the model output power and the actual output power is large, the working time of the generating equipment is expected to increase. In comparison, for operations planning using the PAS predictive data (see Figure 10.13), the generating equipment is never asked to exceed 5 kW in the entire month of December. We see here that operations planning by PAS is at an advantage over planning with past average weather data.

10.6.3.2 Actual System Operation Using the PAS

If a difference occurs in the solar cell output power based on the PAS predictive value and actual weather conditions, the operation of generating equipment

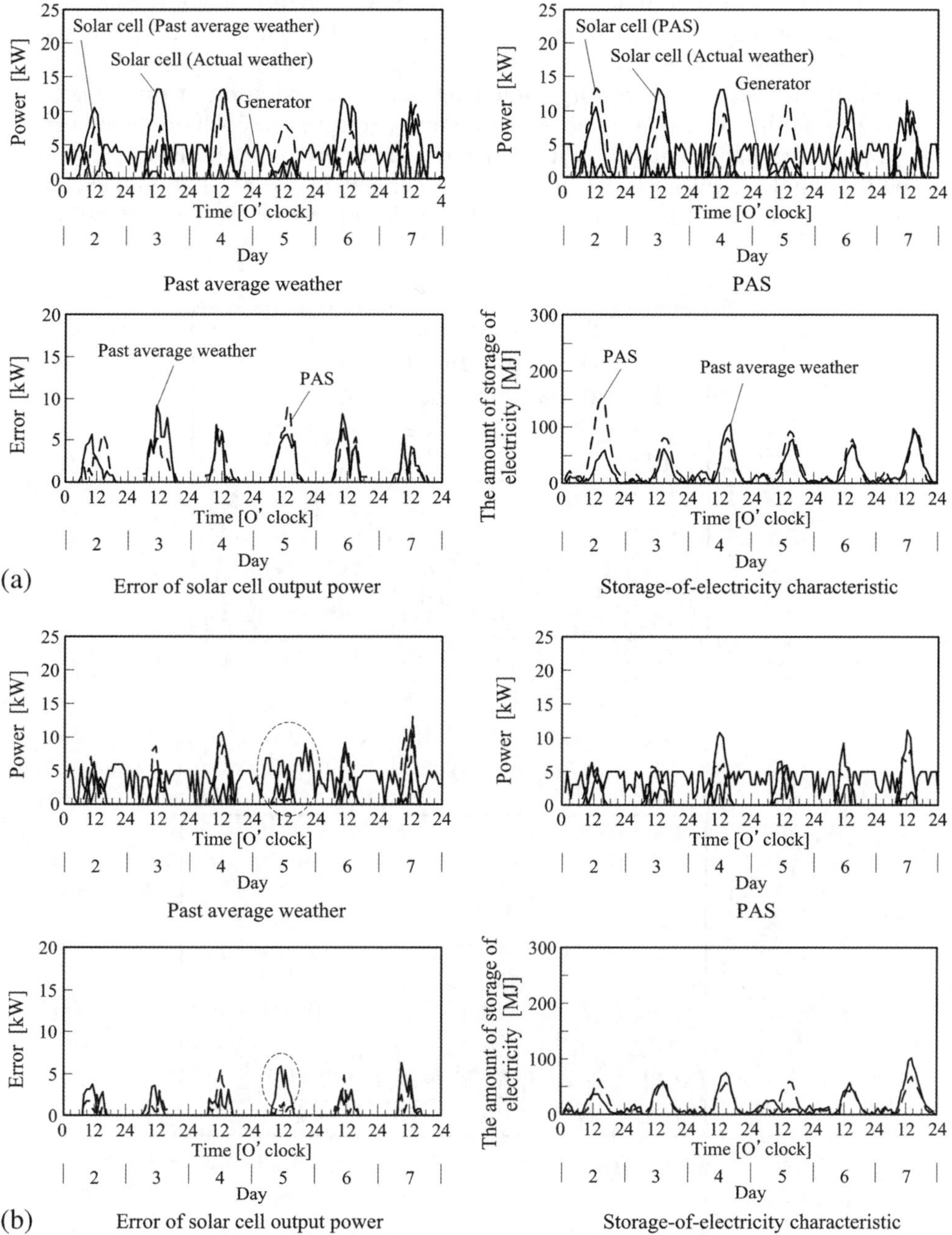

Figure 10.13 Dynamic operation planning based on the past average weather data and the PAS predicted data. In these analyses, the influence on operation planning by the error of "the past average weather data" and "the PAS predictive data" to actual weather is not taken into consideration. (a) March; (b) December.

will follow the method described below. The power balance (eqn (10.2)) of the microgrid in the sampling time t is calculated, and when electricity demand exceeds supply, the generating equipment starts operation immediately. The number of pieces of generating equipment sent into operation at this time is fixed to be the minimum number plausible to avoid the case where the power supply exceeds the demand. When demand is expected to exceed the maximum power supply (5 kW), we can increase the number of pieces of equipment in operation beforehand. Figure 10.14 shows the operation of the generating

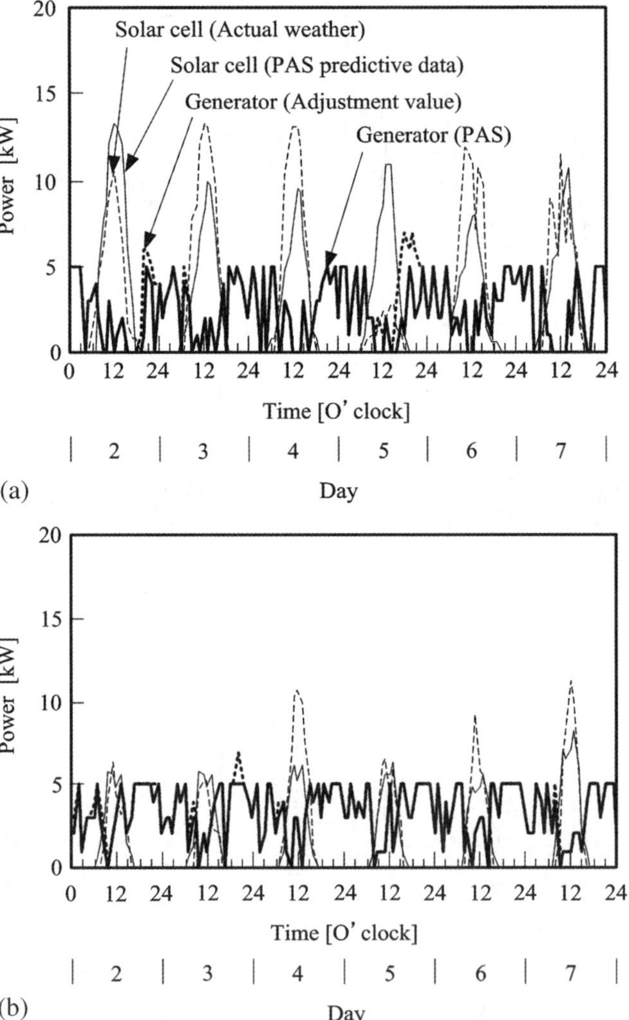

Figure 10.14 Results of the dynamic operation planning of the power generator with PAS prediction. The error of the PAS predictive data to the actual weather is taken into consideration. (a) March; (b) December.

equipment and battery as a result of adding the modification (the operation method described in the above) concerning a prediction error to operation planning using PAS shown in Figure 10.13. Figure 10.14 shows operation planning of the generating equipment using PAS predictive data, as well as operation planning with adjustment of PAS prediction error (here called "Adjustment value"). In actual operation of a microgrid, operation planning with PAS and with "operation planning of the generating equipment by PAS predictive data" and "operation plans by adjustment value" in each sampling time are compared, while a large power output is being performed.

10.6.3.3 Hours of Operation of Generating Equipment

Figure 10.15 shows the result of operating the generating equipment in the microgrid according to the method described in Section 10.6.3.2. When the result of the operation hours of past average weather data and PAS predictive data are compared, the hours of operation based on PAS are lower. The advantageous operation method can be obtained from this rather than the

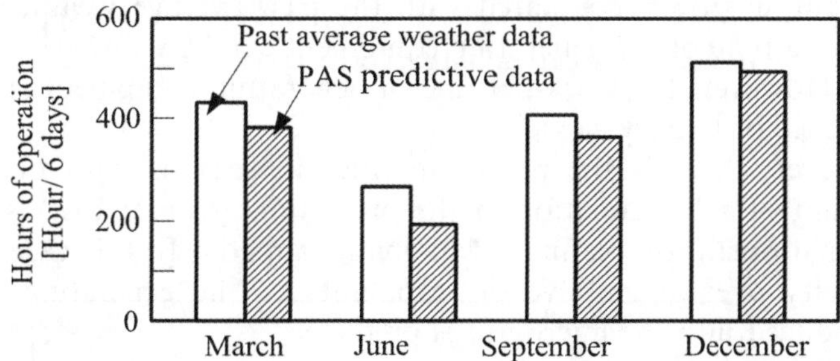

Figure 10.15 Result of the generator hours of operation analyzed under the past average weather data and the PAS predictive data. Consideration of error to the actual value is added.

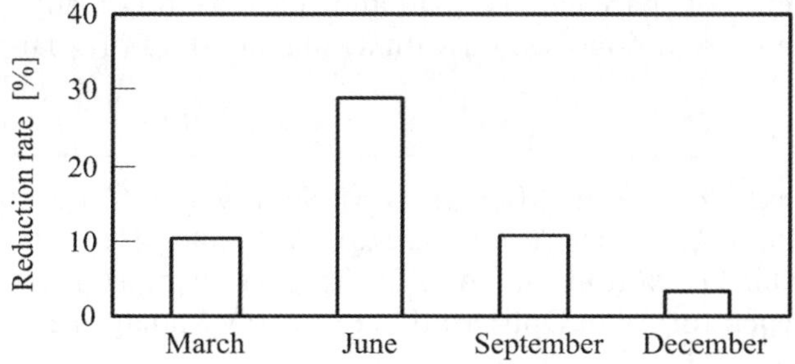

Figure 10.16 Reduction rate of the power generator operation hours in the case of introducing the PAS predictive data to the past average weather data. Consideration of error to the actual value is added.

operation planning method using the past average weather data. Moreover, Figure 10.16 shows the reduction in the total equipment working hours due to using the PAS prediction data rather than using the past average weather data. By introducing PAS into dynamic operation planning of the microgrid, the working time of generating equipment is reduced, as compared to using past average weather data, from 3% to 30%.

10.7 Conclusions

The operations planning of the microgrid was investigated here using a proposed algorithm. Operation of the proposed microgrid was analyzed using actual weather data (amount of solar radiation and outside temperature) collected from the 1st to the 7th in the months March, June, September, and December of 2007 in Sapporo. However, it is thought that the following results change by the introductory region of the microgrid. However, the analysis method described in this chapter can be introduced into various regions. The following conclusions have been obtained:

(1) If the PAS predictive value is introduced as the predictive value of solar cell output power compared with the past average weather data, the working time of the generating equipment can be reduced from 30% to 3%. However, the working time of generating equipment is under the influence of battery capacity.

(2) However, there is the possibility that in seasons with only a small amount of solar radiation, if the prediction error of PAS is large, the original operation planning will change greatly. In this case, a storage capacity over and above the capability of the generating equipment introduced into the microgrid is predicted.

When introducing the proposal analysis method into areas other than Sapporo, it is necessary to investigate the relation between following important factors and system operation plan:

a. influence of space cooling load and space heating load;
b. influence of a rainy season with a small amount of solar radiation.

References

1. Abu-Sharkh, *et al*, Can Microgrids Make a Major Contribution to UK Energy Supply? *Renew. Sustain. Energy Rev.*, 2006, **10**(2), 78–127.
2. M. Muselli, G. Notton and A. Louche, Design of Hybrid-Photovoltaic Power Generator, With Optimization of Energy Management, *Sol. Energy*, 1999, **65**(3), 143–157.
3. Y. Ismail, Y. Kemmoku, H. Takikawa and T. Sakakibara, An Operating Method for Fuel Savings in a Stand-Alone Wind/Diesel/Battery System, *J. Jpn. Sol. Energy Soc.*, 2002, **28**(2), 31–38.

4. Shin'ya Obara, Operating Schedule of a Combined Energy Network System with Fuel Cell, *Int. J. Energy Res.*, 2006, **30**(13), 1055–1073.
5. Shin'ya Obara and Itaru Tanno, Fuel Reduction Effect of the Solar Cell and Diesel Engine Hybrid System with a Prediction Algorithm of Solar Power Generation, *J. Power Energy Syst.*, 2008, **2**(4), 1166–1177.
6. Shin'ya Obara, Dynamic Operation Plan of a Combined Fuel Cell Cogeneration Solar Module and Geo-Thermal Heat Pump System Using Genetic Algorithm, *Int. J. Energy Res.*, 2007, **31**(13), 1275–1291.
7. K. Jorgensen, Technologies for electric, hybrid and hydrogen vehicles: Electricity from renewable energy sources in transport, *Utilities Policy*, 2008, **16**(2), 72–79.
8. Homepage of Japan Meteorological Agency, 2007, http://www.data.jma.go.jp/obd/stats/etrn/index.php.
9. *NEDO Technical information data base*, Standard meteorology and solar radiation data (METPV-3), 2009, http://www.nedo.go.jp/database/index.html.
10. Shin'ya Obara and Itaru Tanno, Operation Prediction of a Bioethanol Solar Reforming System Using a Neural Network, *J. Thermal Sci. Technol.*, 2007, **2**(2), 256–267.
11. K. Narita, *The Research on Unused Energy of the Cold Region City and Utilization for the District Heat and Cooling*, Ph.D. thesis, 1996, Hokkaido University.

CHAPTER 11

Dynamic Characteristics of Power for PEFC Compound System

SHIN'YA OBARA

11.1 Introduction

If small generating equipment is introduced into individual houses and exhaust heat is recovered, the utilization efficiency of the energy will increase greatly compared with existing large-scale power-generation systems. In particular, the fuel cell system is attracting attention because the exhaust gas is clean and the power-generation efficiency is high.[1] Compared with the load fluctuation characteristics by the energy demand side, the dynamic characteristics of a fuel cell system need to be rapid. If the response characteristics of a fuel cell system are slow, the requirement of the demand person cannot be instantly satisfied. As a result, the demand-and-supply balance becomes unstable, and power supply becomes impossible at worst. In order to introduce and apply small-scale fuel cell cogeneration to a building, it is necessary to investigate the response characteristics of the electric power output and the heat power output with load fluctuations. Thus far, the transient response characteristics of the electric power and the heat power of proton-exchange membrane fuel cell (PEFC) cogeneration, with a town-gas reformer and an auxiliary heat source has been investigated.[1] The speed of response of power generation of a cell stack of a PEFC system is very high, and if the control variables of a controller are set up suitably, it is thought, from results of an investigation, that the use of consumer electronics will be satisfactory.[2–5] On the other hand, the speed of

RSC Energy Series No. 3
Compound Energy Systems: Optimal Operation Methods
By Shin'ya Obara and Arif Hepbasli

Published by the Royal Society of Chemistry, www.rsc.org

response of a reformer is slow, compared to a PEFC stack. The response characteristics of the overall system at the time of providing a rapid electric power load to the integrated system by a reformer, a fuel cell stack, an inverter, and an interconnection device, are not known.

Therefore, this chapter considers the determination method for the setup of control variables using a controller, which controls the electric power output of the system. The response characteristics of the electric power, for setting the optimal control variables of proportional action and integral action for the system controller for operating the system, are investigated.[5]

The response characteristics of systems such as time delay and overshooting change by the difference in control variables. Moreover, if it is controlled to follow a system with loads that fluctuate rapidly for a short time, a difference in the load and the response will occur. The magnitude of this difference depends on the fuel consumption of a system, with the response characteristics depending on the control variables. Therefore, it is expected that the response characteristics of a system will change if the control variables are changed, and the amount of consumption of town gas changes as a result. If the control variables of the controller are changed, the transient response characteristics, such as settling time (defined in this chapter as the time for a value to become less than $\pm 5\%$ of a set value), overshoot, rise time, and steady-state error, will change.[6] However, if the control variables are changed, it is expected that the town gas consumption in a system also changes. Therefore, if the control variables are not set to the optimal values, the power-generation efficiency of the system will be low. Moreover, the power load pattern of an individual house is a collection of loads that usually goes up and down rapidly for short periods of time.[6] Thus far, there have been no studies on changes in electric power-generation efficiency when operating a system, wherein the load pattern fluctuates rapidly. Therefore, in this chapter, an electric power load pattern, fluctuating at random for a short time, is introduced into the PEFC cogeneration, and the operation of the system and changes in the power-generation efficiency are clarified. The optimal values of the control variables set up with a controller are also investigated.

11.2 System Description

11.2.1 Outline of System

The PEFC cogeneration system outline examined in this chapter is shown in Figure 11.1. The town gas supplied to the system includes a quantity that is changed into reforming gas by a reformer ($Q_{rm,t}$), and a quantity consumed by a burner as an object for the heat source of a reforming reaction ($Q_{cb,t}$). The relationship between the supply, the sum of the two town gas quantities ($Q_{rm,t} + Q_{cb,t}$), and the electric energy output of the system ($E_{sys,t}$) is shown in Figure 11.2.[7,8] The calorific power of hydrogen in the reforming gas generated by the reformer divided by the supply quantity of heat of the town gas is defined

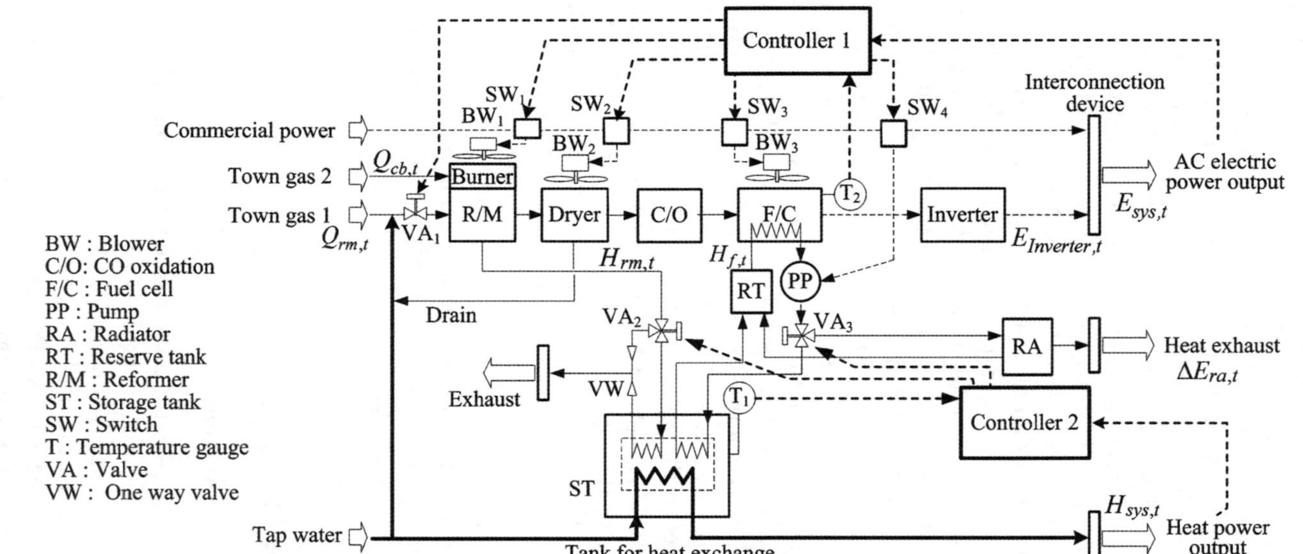

Figure 11.1 Fuel cell cogeneration system for individual houses.

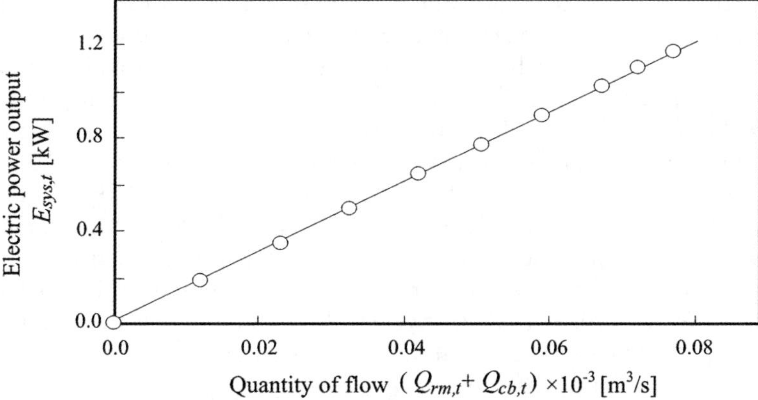

Figure 11.2 The power output of the system outlet. The anode and cathode electrode areas of the fuel cell stack are 0.5 m^2, respectively. The reformer efficiency is 73%.

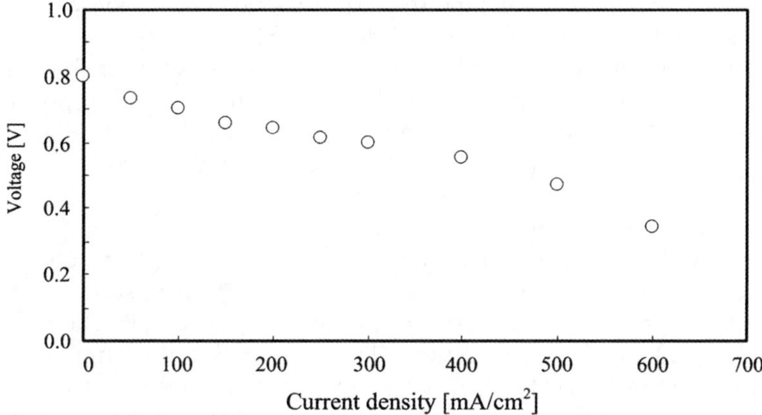

Figure 11.3 Single-cell performance generated with reformed gas and air. Operating temperature 333 K, and reactant flow stoichiometries 2.0 both hydrogen and oxygen.

as the reformer efficiency $\eta_{rm,t}$ at sampling time t. $\eta_{rm,t}$ is set at 73% in this system.[9] The electrode areas of the anode and the cathode in the fuel cell stack are set to 0.5 m^2, respectively, and the output characteristics of a single cell in the fuel cell assumed in this chapter is shown in Figure 11.3.[7,8] The maximum power-generation efficiency of the fuel cell stack is 48%.

(a) Efficiency of a carbon-monoxide oxidization system and an inverter.

A significant quantity of water is present in the reforming gas produced by the reformer. Therefore, the reforming gas is cooled by an air supply from a blower with a dryer system and condensation separation of the water is carried out. In order for the carbon-monoxide concentration of the reforming gas in the fuel cell stack inlet to be 10 ppm or less, a carbon-monoxide oxidization system is prepared. Although carbon monoxide is burned using a catalyst and converted into carbon dioxide in the carbon-monoxide oxidization system, it is assumed in this chapter that nearly 2% of hydrogen in the reforming gas will also be burned. Therefore, the efficiency of the carbon-monoxide oxidization system in this chapter is 98%. Reforming gas is supplied to the fuel cell stack from the carbon-monoxide oxidization system, generating electricity, and the generated DC electric power is changed into AC power of a fixed frequency using an inverter, which is then supplied to an interconnection device. The efficiency of the inverter was set at 95% in this chapter.

(b) The heat supply method.

The exhaust heat of the reformer and the fuel cell stack is supplied to a heat-storage tank (ST), and the heat of tap water is exchanged for the heat-storage medium with a heat exchanger installed in the heat-storage tank, which then supplies hot water to the demand side. The thermal storage medium is contained in the heat-storage tank. When the exhaust heat exceeds the heat demand, the heat-transfer medium is heated and the heat is stored. The response time of the exhaust heat output in the reformer and the fuel cell stack is long compared with the response time of the heat output from thermal storage.[10] Therefore, in order to obtain a quick response, the heat in the thermal storage is given priority and used first. When the amount of exhaust heat exceeds the heat demand, it is necessary to release the heat. In this case, valves VA_2 and VA_3 are operated and the excess heat is released from the system.

(c) The controller and auxiliary machinery.

This system has two controllers, as shown in Figure 11.1. Controller 1 is for the power-generation system, and Controller 2 is for the heat supply system. In each controller, the control variable for proportional action (*P* action), integral action (*I* action) and derivative action (*D* action) can be set up. In Controller 1, the temperature of the fuel cell stack and the amount of electricity demand are the input data, and in Controller 2, the heat-transfer-medium temperature of the heat-storage tank and the heat amount demand are the input data. The electric power generated by the fuel cell system and commercial electric power can be supplied to the demand side through an interconnection device. The auxiliary machinery of the system is operated with commercial electric power. The auxiliary machinery includes an exhaust heat-circulating pump (PP) for the fuel cell stack, a blower for the reformer burners (BW_1), a blower for the dryers (BW_2) and a blower for the fuel cell cathode (BW_3). By using commands from Controller 1 and Controller 2, the switches SW_1, SW_2, SW_3, and SW_4 are operated, and the starting and stopping of the auxiliary machinery are performed.

11.2.2 System-Control Block Diagram

Figure 11.4 shows the control block diagram of the system shown in Figure 11.1. The time constants for each piece of equipment shown in this figure are

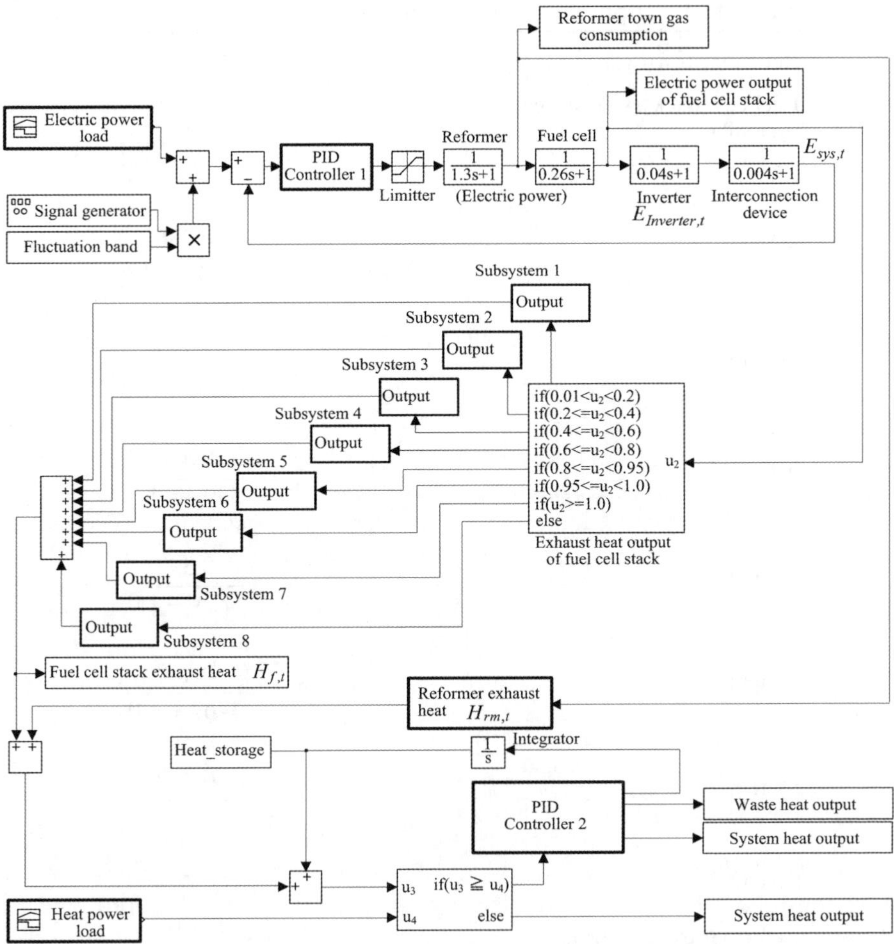

Figure 11.4 System control block diagram.

determined by test evaluation data and reference data, the details of which are given in ref. 9. The input values used in the analysis are shown in Table 11.1. However, the power load and heat load are given in a step input. Investigation of the transient response characteristics can be performed by providing an electric power load and a heat power load to the system shown in Figure 11.4. For the electric power load, random changes can be added for every sampling time. For each controller, limiters are included so that extremely large overshoots are not generated for control by Controller 1 and Controller 2. Subsystems 1 through 8 represent the response characteristics of the exhaust heat output in a fuel cell stack, and details are shown in Figure 11.5. The parameters *a* and *b* in this figure express the time constant and the constant part of a primary delay system transfer function, respectively. The exhaust heat of the fuel cell stack serves as a nonlinear output to the electric power load.[9] In this

Table 11.1 Input data used in analysis.

Reformer town gas consumption	$0.0659.\ E_{sys,t}\ m^3/s$
Fuel cell stack exhaust heat	
System 1	$\dfrac{0.123}{1517s + 1.0}$
System 2	$\dfrac{0.236}{1374s + 1.0}$
System 3	$\dfrac{0.327}{1163s + 1.0}$
System 4	$\dfrac{0.438}{599s + 1.0}$
System 5	$\dfrac{0.397}{887s + 1.0}$
System 6	$\dfrac{0.460}{346s + 1.0}$
System 7	$\dfrac{0.469}{133s + 1.0}$
System 8	0
Reformer exhaust heat	$\dfrac{1.0}{33.3s + 1.0}$
Limitter	0 to 1.5 kW

study, the relationship between electric power load and the exhaust heat output of a fuel cell is divided into eight zones. The relationship between the electric power load in each zone and the exhaust heat output are approximated by the linear formulas expressed with Subsystem 1 to Subsystem 8.

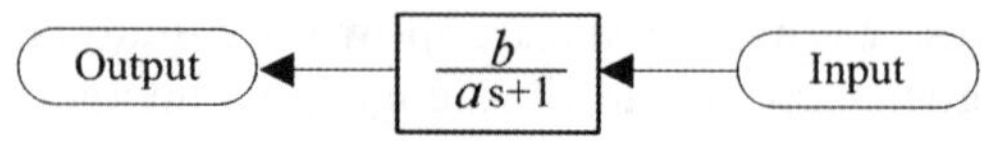

Figure 11.5 Block diagram of the Subsystems from 1 to 8.

11.2.3 The Analysis Method

The transient response characteristics of the electric power output in this system are analyzed using MATLAB (Ver.7.0)/Simulink (Ver.6.0) of Math Work. The Runge–Kutta method was used as the numerical method.[11–13] The sampling time of the analysis was calculated automatically and the analysis error is less than several per cent.

11.3 System Control

11.3.1 The Input of the System

The operational model of the system shown in Figure 11.1 is shown in Figure 11.6. Figures 11.6(a) and (b) are the models that show the electric power load input and the heat power load input, respectively, and both power loads are zero between the sampling times t_1 and t_0. At sampling time t_1, although the step input of the electric power load and the heat power load is carried out in the system, the heat power load is fixed until t_3, and a still higher step input is given at t_2 for the electric power load. The electric power load in individual houses fluctuates within a short period. Therefore, in the analysis in the following section, random load fluctuations are given for every sampling time in $\pm 5\%$ and $\pm 10\%$ intervals of the range of the electric power load input into the system. For the analyses shown in Figures 11.7–11.11 described in the following section, the time from t_0 to t_1 is 5 s, the time from t_1 to t_3 is 20 s, and the time from t_3 to t_4 is 5 s.

11.3.2 Control of Startup

Figure 11.6(c) shows the model of the switches and the valves of the system operated by the controller of the operation. The symbols in this figure correspond to the symbols shown in Figure 11.1. During the startup of the system, air is supplied to the burner, which is the heat source for the reformer and the dryer system. For this, switches SW_1 and SW_2 are turned ON by a command from Controller 1, and blowers BW_1 and BW_2 are operated. At sampling time t_1, commands are given to SW_3 by Controller 1, and BW_3, which supplies air to the cathode of the fuel cell stack, is operated. Simultaneously, a control command is given by Controller 1 to VA_1 so that valve VA_1 can be opened. As shown in Figure 11.6(d), operation of the fuel cell system follows the electric power load. This depends, for the response characteristics of the electric power, on transient response characteristics, such as that of the reformer, the fuel cell stack, and the inverter. Therefore, compared with the characteristics of the input shown in Figure 11.6(a), time delay, overshoot and steady-state error cause the response characteristics shown in Figure 11.6(d).

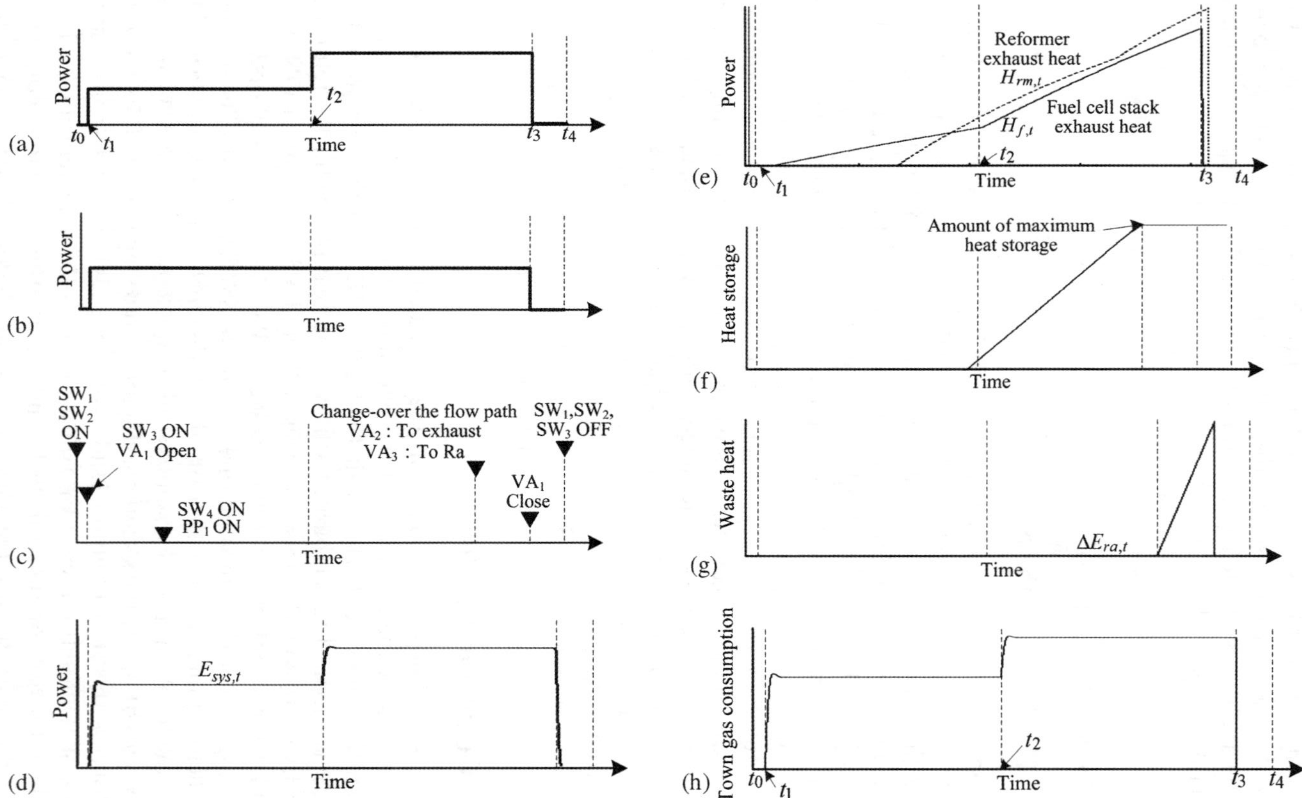

Figure 11.6 System operation model; (a) Electric power load; (b) Heat power load; (c) Control system operation; (d) System electric power output; (e) Heat power output; (f) Amount of heat storage; (g) Amount of waste heat from the system; (h) Consumption of town gas.

11.3.3 Control of Heat Output

Since the speed of response of the heat exhaust output from the fuel cell stack and the reformer is slow, heat output takes time. If the temperature T_2 of the fuel cell stack rises, SW_4 will be turned ON by an order from Controller 1, and the heat exhaust circulating pump PP will be operated. Figure 11.6(e) shows the model for this operation. Furthermore, when the fuel cell stack and the reformer of heat exhausts exceed the heat amount demand, as shown in Figure 11.6(f), thermal storage of the excessive heat is carried out using a heat-storage tank. Finally, when the heat exhaust exceeds the thermal storage capacity of the heat-storage tank, valves VA_2 and VA_3 are operated by Controller 2, and as shown in Figure 11.6(g), surplus heat is released from the system.

11.3.4 Town-Gas Consumption

Figure 11.6(h) shows the model for town gas consumption in the system. Since town gas consumption is dependent on the amount of electric power generated by the fuel cell, the transient characteristics of the town gas consumption are the same as shown in Figure 11.6(d). With control variables set up by Controller 1, the output characteristics of Figure 11.6(d) differ greatly. Therefore, the town gas consumption in the system also changes with the value of the control variables set up by Controller 1.

11.4 Results and Discussion

11.4.1 Control Variables and the Response

The analysis results of the response characteristics of control variables and electric power set up by Controller 1 are shown in Figure 11.7. Figures 11.7(a)–(c) show the analysis results of the response characteristics due to the difference in the control variable of *P* action and *I* action under each electric power load. When *P* action is independently used to control this system, the steady-state error is large. Moreover, for the use of the *D* action, such as *D* control, *PD* control, and *PID* control, the response characteristics of the system become extremely unstable. Therefore, in this analysis, independent control by *P* action and control by the *D* action are not used. The results in Figures 11.7(a)–(c) show that the optimal values of the control variable used by Controller 1 differ in the magnitude of the electric power load. Therefore, if the control variables of the system controller are changed with the magnitude of electric power load, it is predicted that the transient response characteristics of the system will become favorable.

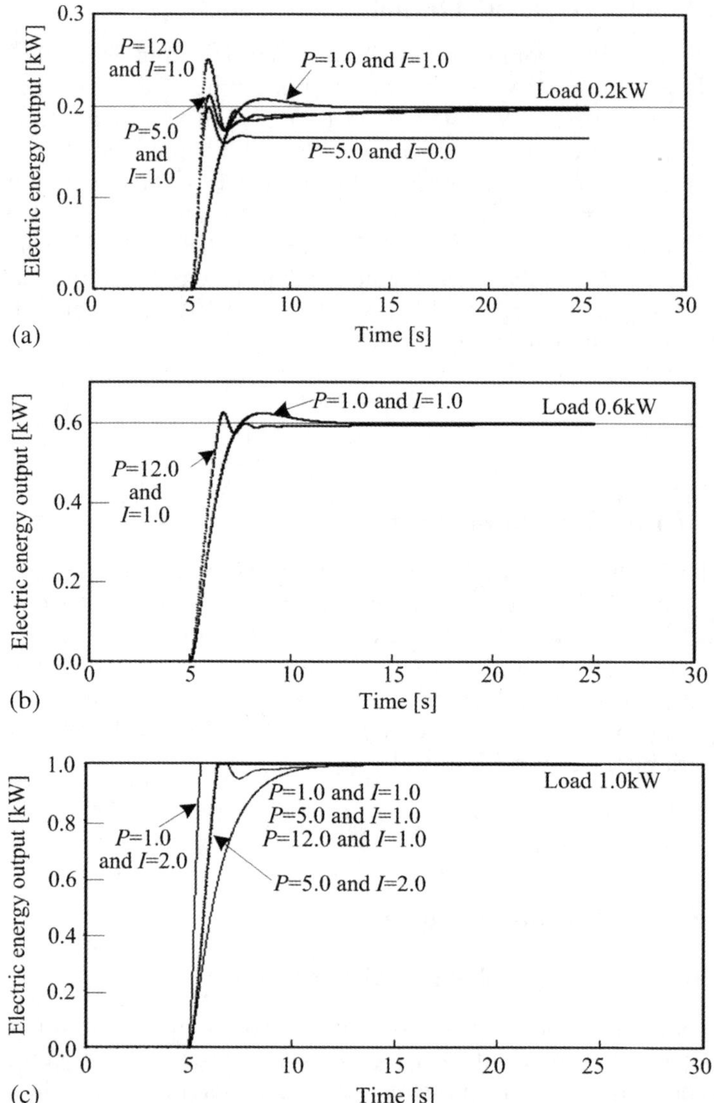

Figure 11.7 Characteristics of electric power output of the system; (a) Electric power load is 0.2 kW; (b) Electric power load is 0.6 kW; (c) Electric power load is 1.0 kW.

11.4.2 The Response Characteristics of the System

(1) The amount of electric power output.

If a step load is provided to the fuel cell system, until the operation of the system reaches a settling time, the electric power outputs will go up and down along with the load. The amount of integration of the electric power load for a definite period of time (this is written below as the amount of theoretical electric power loading) differs from the electric energy output by the system, because of the upper and lower sides of the electric power output. Figure 11.8 shows the calculation results of the difference in the amount of theoretical electric power loading and the amount of system electric power outputs (*err*), when adding electric power load with fluctuations shown in this figure to the system for 20 s. The results of $\pm 5\%$ and $\pm 10\%$ of the load fluctuation rate in Figures 11.8(a)–(c) are not shown for the zone where *P* operation is large. The reason for this is that the calculation results diverge in the zone where the *P* operation is large. This tendency is strong as there is lower electric power output from the system than the amount of theoretical electric power loading, and the rate of load fluctuation becomes large, by all the results shown in Figure 11.8. A settling time, as shown in Figure 11.7(a) to (c), is five seconds. Also, the electric power outputs are less than the value of electric power load in many cases. Moreover, the control variable set up with a controller has a large influence on the rising time of a response. For this reason, the results shown in Figure 11.8 of the amount of electric power output in the system are also smaller than the amount of theoretical electric power loading as the rate of load fluctuation becomes large.

(2) Electric power-generation efficiency.

The electric power-generation efficiency of the system can be calculated from the electric energy generated by the system by dividing it by town gas consumption. However, the response characteristics of the electric power output from the system do not satisfy the amount of theoretical electric power loading, as Figure 11.8 showed. Therefore, the following equation (defines the electric power-generation efficiency η_e of the system in this chapter,

$$\eta_e = \left(E_t / V_g\right) + err \tag{11.1}$$

Although the cold start of a reformer takes about 1 h,[14] in this analysis, a reformer changes into the condition of already having completed the warm-up. Therefore, the quantity of gas used for the setup operation of the reformer is not included in the town gas consumption, V_g. The control variables of *P* action and *I* action, and the relationship of η_e are calculated for the rate of load fluctuations of the electric power load for different electric power loads, and these results are shown in Figure 11.9. The maximum electric power-generation efficiency of this system is 33%. For each load, except for the 1.0 kW electric power load, the system can be operated with a high value of η_e by making the control variable of *I* action about 1 to 2, and by making the control variable of *P* action about 5 or less. For 1.0 kW of electric power load, the operation of the system at the high value of η_e of the control variable zones of *P* action and *I* action is narrow, ranging from 1 to 2 for the control variable of *I* action, and 1 or less for the control variable of *P* action. For this reason, the maximum load of the system is 1 kW, and it is possible that the zones of the optimal control variables set up with a controller are limited.

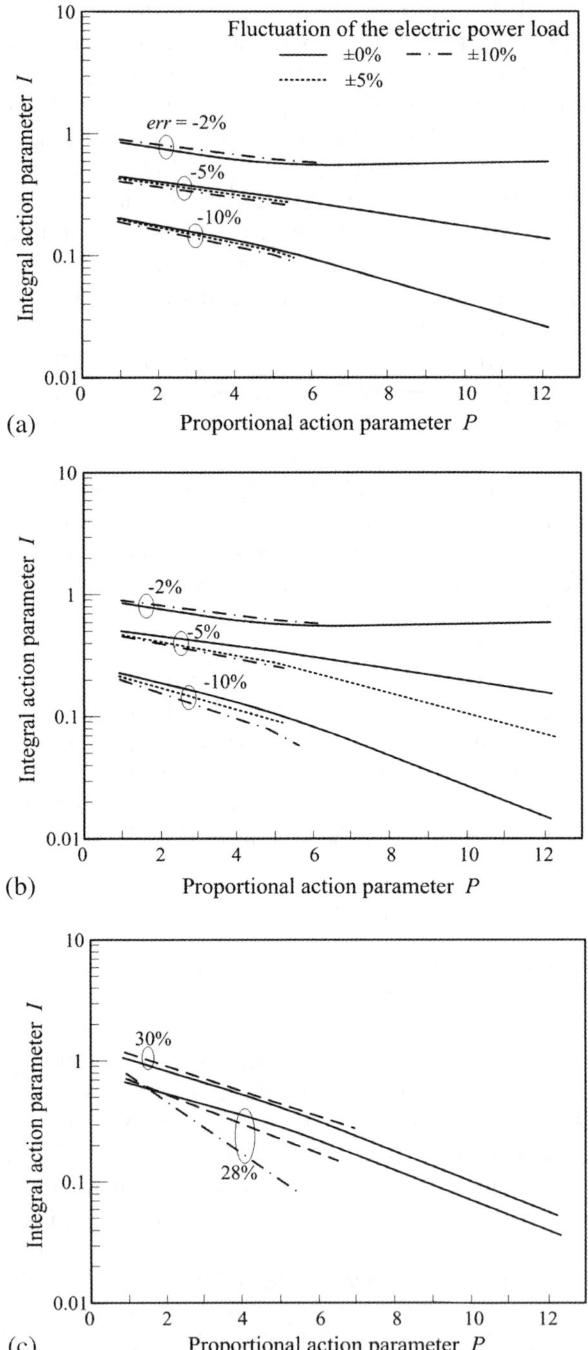

Figure 11.8 Electric power generation error for the system; (a) Electric power load is 0.2 kW; (b) Electric power load is 0.5 kW; (c) Electric power load is 1.0 kW.

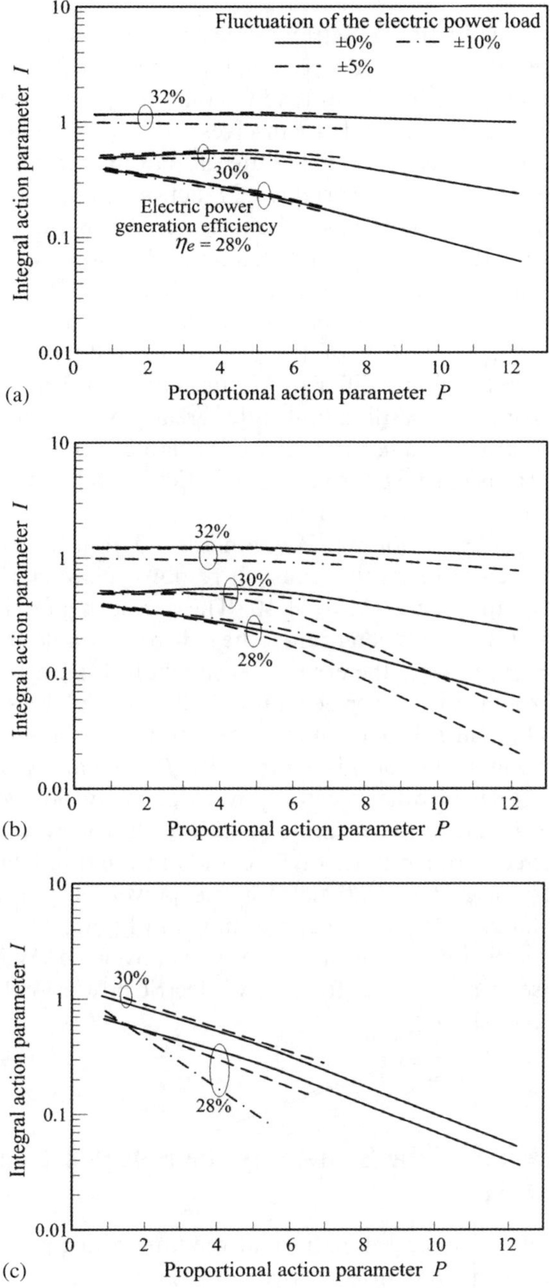

Figure 11.9 Electric power-generation efficiency η_e of the system; (a) Electric power load is 0.2 kW; (b) Electric power load is 0.5 kW; (c) Electric power load is 1.0 kW.

(3) Settling time.

The calculation results for the settling time are shown in Figure 11.10. In the analysis results of settling time, the reasonable zones of the control variables *P* and *I* change greatly with the rate of load fluctuations of the electric power load as well as the electric power loads themselves. Although a reasonable settling time is obtained for the case where the rate of load fluctuation is 0 for 0.2 kW electric power load, the control variable of *P* action is limited to nearly 5, and the control variable of *I* action is a large zone. When the power loads are 0.5 kW and 1.0 kW, the control variable zone of *I* action by each load is 3 or more. For the results of the settling time in $\pm 5\%$ rate of load fluctuation, any power loads have the same value and these zones of a control variable are also the same. However, at $\pm 10\%$ of the rate of load fluctuation, the control variable zones of settling time differ in the amount of electric power load. The control variable zones of a settling time differ greatly by electric power load and rate of load fluctuation because the control variables for set up have a strong influence on the transient response characteristics of the system.

(4) Selection of control variables.

The control variables of optimal *P* action and *I* action for every power load are chosen from each result of the transient response characteristics of the fuel cell system shown in Figures 11.8–11.10. These results are shown in Figure 11.11. The reason for having determined the control variable zone in 0.2 kW of power loads in Figure 11.11, the error (*err*) shown in Figure 11.8(a) is small in $I = 0.7$ or more zones, the power-generation efficiency of the system shown in Figure 11.9(a) is high in $I = 1.0$ or more zones, and also, the settling time of the system shown in Figure 11.10(a) is a zone with $P = 5.0$ and $I = 2.0$ to 7.0. The control variable zones in other electric power loads considered to be optimal were determined similar to the example of 0.2 kW electric power load described above. Using the control variables of *P* action and *I* action from Figure 11.11, the transient response characteristics for the power output of the fuel cell system are calculated, and the results are shown in Figure 11.12. In the system controller, the electric power output in the system over 1.0 kW is restricted, the transient response characteristics for 1.0 kW electric power load in Figure 11.12 do not have an overshoot.

11.4.3 Operation of the System by the Selected Control Variables

The simulation of the operation pattern at the time of the power load input in Figure 11.6(a) into the system is carried out using the control variables of *P* action and *I* action shown in Figure 11.11. In Figure 11.6(a), the time t_0 to t_1 is 5 s, time t_1 to t_2 and time t_2 to t_3 are 100 s, respectively, and time t_3 to t_4 is 5 s. The reason for the period of t_1 to t_3 being 200 s is that the time period needs to be sufficient in order to perform observation of the transient response characteristics of the system, and because the differences in the responses of the

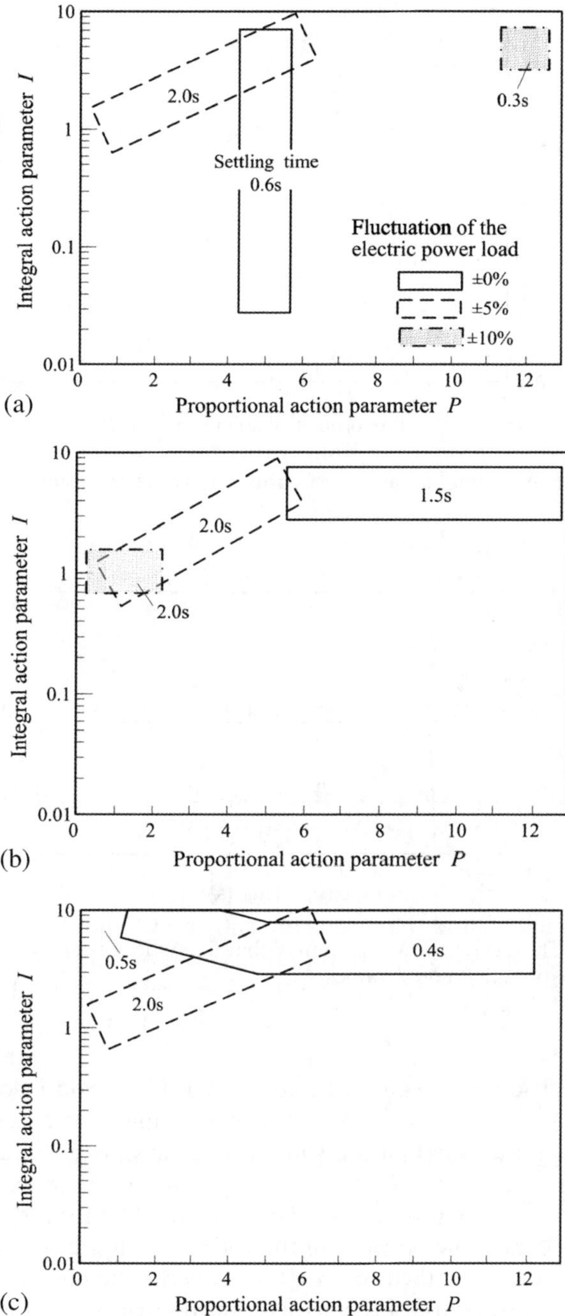

Figure 11.10 The settling time for electric power generation for the system; (a) Electric power load is 0.2 kW; (b) Electric power load is 0.5 kW; (c) Electric power load is 1.0 kW.

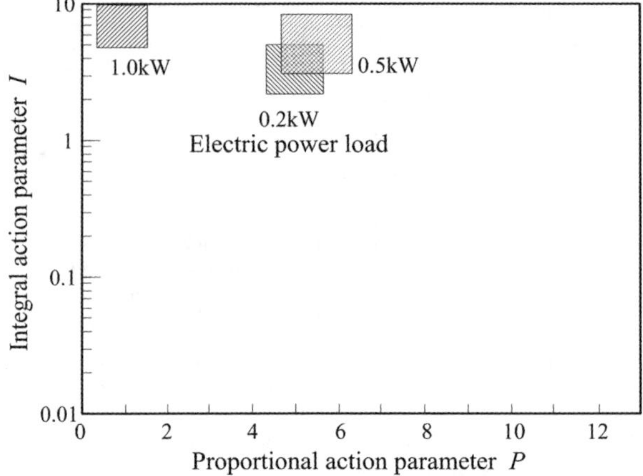

Figure 11.11 The optimal areas for PI control for different electric power loads.

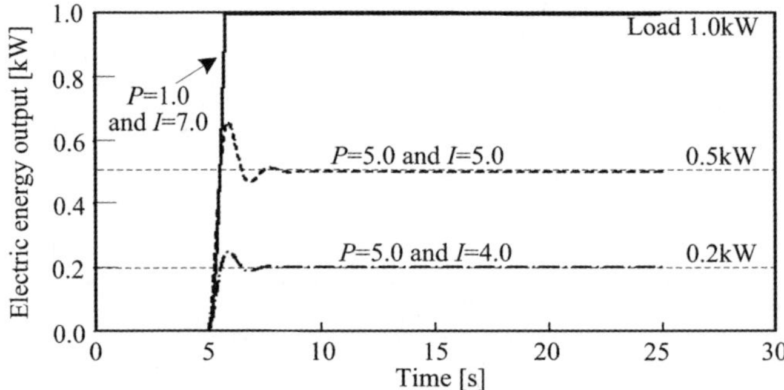

Figure 11.12 The electric power response characteristics of the system analyzed with optimal *P* and *I* values.

amounts of electric power cannot be compared if a period is long. Moreover, time t_1 to t_2 is 0.2 kW of electric power load and time t_2 to t_3 is 0.5 kW. Figure 11.13 shows the power load patterns input into the system. Figure 11.14 shows the calculation results of the electric power output response of the system corresponding to the load inputs in Figure 11.13. The results of the response characteristics of the power output of the system in Figures 11.14(a)–(c) are the same. Figure 11.15 shows the response results of the exhaust heat output by the fuel cell stack and the reformer. The exhaust heat output of the fuel cell stack and the reformer shows a late speed of response. Therefore, the heat exhaust output of the fuel cell stack and the reformer requires progress of time from starting of the system. As one method of improving the speed of response of the heat, the period shift of the heat supply-and-demand by thermal storage is

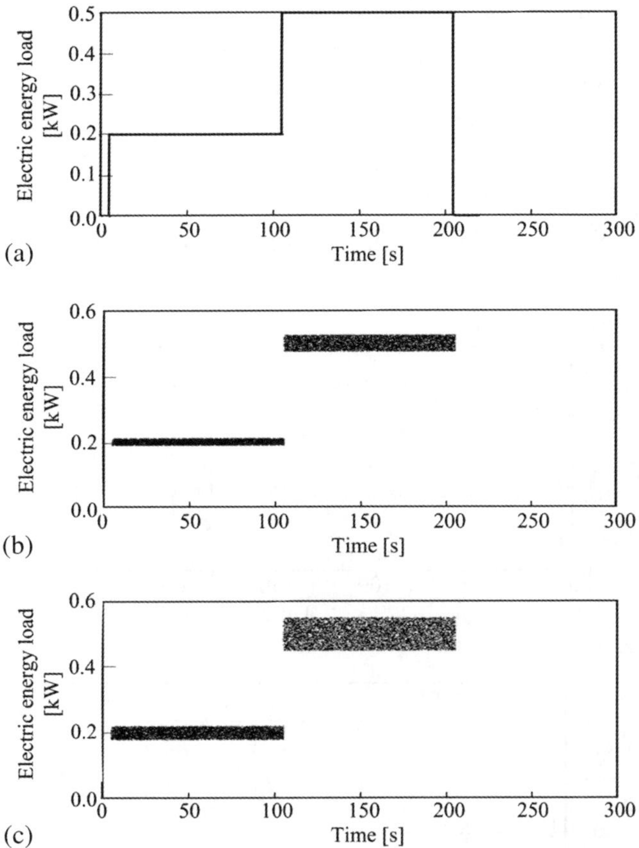

Figure 11.13 Electric power load patterns input into the system; (a) Electric power load fluctuation is ±0%; (b) Electric power load fluctuation is ±5%; (c) Electric power load fluctuation is ±10%.

effective. Figure 11.16 shows the calculation results of the amount of town gas consumed by the system. The consumption of town gas is hardly influenced by the rate of load fluctuations of the electric power like the response results of the electric power shown in Figure 11.14. If the control variables set up with the controller of the fuel cell system are able to be set up by the results in Figure 11.11 according to the magnitude of electric power load, the characteristics of transient response of a settling time, the steady-state error, the overshoot, and the rising time can be changed to the best conditions, and the system can be operated. Moreover, by changing the control variables set up with the controller of the fuel cell system for the magnitude of electric power loads, town gas consumption and electric power-generation efficiency are hardly influenced by additions of fluctuations of the electric power load to the system over a short period.

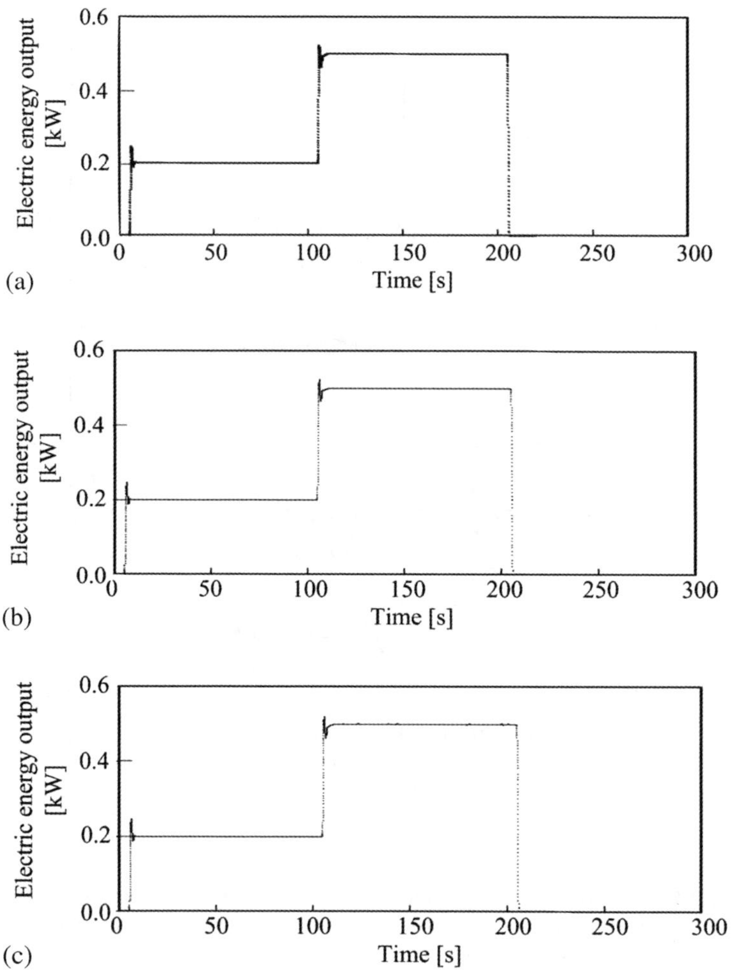

Figure 11.14 Response characteristics of the electric power output of the system; (a) Electric power load fluctuation is $\pm 0\%$; (b) Electric power load fluctuation is $\pm 5\%$; (c) Electric power load fluctuation is $\pm 10\%$.

11.5 Conclusions

The determination method of the control variables of a system controller, when introducing and carrying out the load-following operation of the PEFC cogeneration to the electric power loads that randomly go up and down, for individual houses was considered. In particular, since the electric power load pattern of an individual house consists of loads that often rapidly rise and fall in a short time, the difference in the load and the system response is influenced by the control variable. So, in this chapter, the relation of load with fluctuations, control variables, and settling time was investigated by numerical analysis. As a result, this chapter showed that the operation was minimally

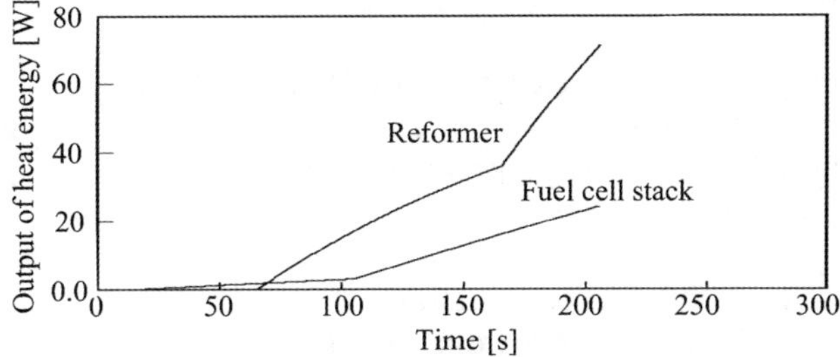

Figure 11.15 Heat power output of the fuel cell stack and the reformer. Electric power load fluctuations are ±0%, ±5%, and ±10%.

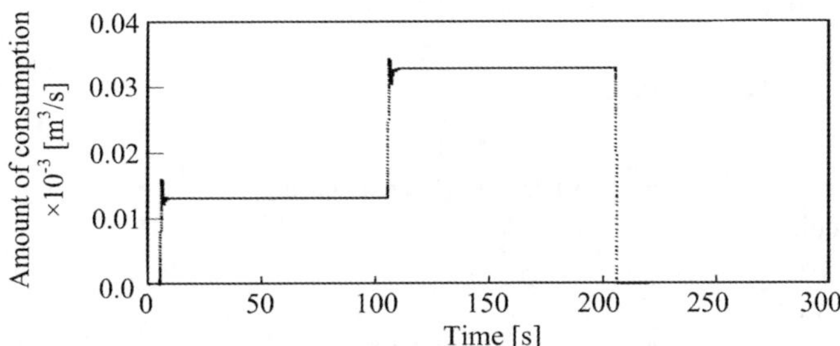

Figure 11.16 Town gas consumption ($Q_{rm,t} + Q_{cb,t}$) as a function of time. Electric power load fluctuations are ±0%, ±5%, and ±10%.

influenced by load fluctuations and could be performed by changing control variables with the magnitude of the electric power loads added to the system. If the control variables of the system controller are changeable with the magnitude of electric power load, even if a change in the electric power load in a short period is added to a system, town gas consumption and the power-generation efficiency are minimally influenced and can continue operation. Moreover, in order to maintain the power-generation efficiency of the system, it was shown that the selected ranges of the control variable of P and I operation must be very narrow.

Acknowledgments

This work was partially supported by a Grant-in-Aid for Scientific Research (C) from the JSPS.KAKENHI (15510078).

Nomenclature

a: time constant in a primary delay system transfer function s
b: constant part in a primary delay system transfer function

D: control variable of derivative action
E: electric power kW
E_t: amount of theoretical electric power loading J
err: difference of amount of theoretical electric power loading and the amount of system electric power outputs %
H: heat power kW
ΔH: waste heat power kW
I: control variable of integral action
P: control variable of proportional action
Q: quantity of flow m^3/s
t: sampling time s
V_g: heat quantity of town gas consumption J
η_e: electric power-generation efficiency %
$\eta_{rm,t}$: reformer efficiency %

Equipment Symbols

cb: burner for the heat sources of the reformer
f: fuel cell
ra: radiator
rm: reformer
sys: system

References

1. S. Obara and K. Kudo, Study on Small-Scale Fuel Cell Cogeneration System with Methanol Steam Reforming Considering Partial Load and Load Fluctuation, *Trans. ASME, J. Energy Res. Technol.*, 2005, **127**, 265–271.
2. J. Hamelin, K. Agbossou, A. Laperriere, F. Laurencelle and T. K. Bose, Dynamic behavior of a PEFC stack for stationary applications, *I. J. Hydrogen Energy*, 2001, **26**, 625–629.
3. C. J. Hatziadoniu, A Simplified Dynamic Model of Grid-Connected Fuel-Cell Generators, *IEEE Trans. Power Deliv.*, 2002, **17**(2), 467–473.
4. S. Kourosh, Dynamic and Transient Analysis of Power Distribution System with Fuel Cells. Part 1: Fuel Cell Dynamic Model, *IEEE Trans. Conversion*, 2004, **19**(2), 423–428.
5. G. J. Silva, A. Datta and S. P. Bhattacharyya, *PID Controllers for Time-Delay Systems*, 2003, Birkhauser, Boston, USA.
6. S. Kurata, Measurement of a domestic energy consumption pattern, *National Institute of Advanced Industrial Science and Technology*, News No.12, 1999, 1–14 (in Japanese).
7. Ibaraki Prefecture Government Office of Education, Modeling of hydrogen energy system, *High school active science project research report*, 2002, Ibaraki, Japan (in Japanese).

8. M. Mikkola, *Experimental Studies on Polymer Electrolyte Membrane Fuel Cell Stacks*, Master's thesis submitted in partial fulfillment of the requirements for the degree of Master of Science in Technology, 2001, Helsinki University of Technology, 58–79.
9. S. Obara and K. Kudo, Relation between Control Variables of PEFC System and Power-Generation Efficiency, *Trans. Jpn. Soc. Mech. Eng. B*, 2005, **72**(714), 447–454 (in Japanese).
10. S. O. Morner and S. A. Klein, Experimental Evaluation of the Dynamic Behavior of an Air-Breathing Fuel Cell Stack, *ASME J. Sol. Energy Engineering*, 2001, **123**(3), 225–231.
11. J. R. Dormand and P. J. Prince, A family of embedded Runge–Kutta formulae, *J. Comput. Appl. Math.*, 1980, **6**, 19–26.
12. G. Forsythe, M. Malcolm and C. Moler, *Computer Methods for Mathematical Computations,* 1977, Prentice-Hall, New Jersey.
13. D. Kahaner, C. Moler and S. Nash, *Numerical Methods and Software,* 1989, Prentice-Hall, New Jersey.
14. Y. Zhang, M. Ouyang, Q. Lu, J. Luo and X. Li, A Model Predicting Performance of Proton Exchange Membrane Fuel Cell Stack Thermal Systems, *Appl. Thermal Eng.*, 2004, **24**, 501–513.

CHAPTER 12

Performance Analysis and Assessment of Compound Energy Systems Using Exergy Analysis Method

ARIF HEPBASLI

12.1 Introduction

Nowadays, with the increase in the world population and industrialization, there has been a continuous increase in the consumption of fossil fuels, which further causes several problems such as increased carbon emissions mainly from fossil-fuel combustion products and leads to global problems. In this regard, the world has sounded the alarm "accelerated global warming", while the need for clean and renewable energy has become inevitable.[1]

The European Union (EU) from the beginning of 2007 has focused its emphasis on developing a new policy that puts energy back at the heart of EU action.[2] European energy strategy and policy are strongly driven by the twin objectives of sustainability (including environmental aspects) and security of supply. Implementation of new energy technologies is of major importance to satisfy these objectives, such as renewable energy systems at the supply side and improved energy end-use efficiency at the demand side.[2,3]

Research into future alternatives has been and is still being conducted with the aim of solving the complex problems of this recent time, *e.g.*, rising energy requirements of a rapidly growing world population and global environmental

RSC Energy Series No. 3
Compound Energy Systems: Optimal Operation Methods
By Shin'ya Obara and Arif Hepbasli

Published by the Royal Society of Chemistry, www.rsc.org

pollution. Therefore, options for a long-term and environmentally friendly energy supply have to be developed leading to the utilization of renewable energy resources (*i.e.*, water, sun, wind, biomass, geothermal and hydrogen) and fuel cells. Renewables could shield a nation from the negative effect in the energy supply, price and related environment concerns.[4]

Dincer[5] reported the linkages between energy and exergy, exergy and the environment, energy and sustainable development, and energy policy making and exergy in detail. He provided the following key points to highlight the importance of the exergy and its essential utilization in numerous ways: (a) it is a primary tool in best addressing the impact of energy resource utilization on the environment. (b) It is an effective method using the conservation of mass and conservation of energy principles together with the second law of thermodynamics for the design and analysis of energy systems. (c) It is a suitable technique for furthering the goal of more efficient energy-resource use, for it enables the locations, types, and true magnitudes of wastes and losses to be determined. (d) It is an efficient technique revealing whether or not and by how much it is possible to design more efficient energy systems by reducing the inefficiencies in existing systems. (e) It is a key component in obtaining a sustainable development.

Sustainable development does not make the world 'ready' for the future generations, but it establishes a basis on which the future world can be built. A sustainable energy system may be regarded as a cost-efficient, reliable, and environmentally friendly energy system that effectively utilizes local resources and networks. It is not 'slow and inert' like a conventional energy system, but it is flexible in terms of new technoeconomic and political solutions. The introduction of new solutions is also actively promoted.[6,7]

Exergy, which is a means to sustainable development, has become a very effective tool in evaluating system effectiveness and in designing the system to maximize energy savings.[8] The concepts of exergy, available energy, and availability are essentially similar. The concepts of exergy destruction, exergy consumption, irreversibility, and lost work are also essentially similar. Exergy is also a measure of the maximum useful work that can be done by a system interacting with an environment that is at a constant pressure P_0 and a temperature T_0. The simplest case to consider is that of a reservoir with heat source of infinite capacity and invariable temperature T_0. It has been considered that maximum efficiency of heat withdrawal from a reservoir that can be converted into work is the Carnot efficiency.[9]

For providing an efficient and effective use of fuels, it is essential to consider the quality and quantity of the energy used to achieve a given objective. In this regard, the first law of thermodynamics deals with the quantity of energy and asserts that energy cannot be created or destroyed, whereas the second law of thermodynamics deals with the quality of energy, *i.e.*, it is concerned with the quality of energy to cause change, degradation of energy during a process, entropy generation and the lost opportunities to do work. More specifically, the first law of thermodynamics is concerned only with the magnitude of energy with no regard to its quality; on the other hand, the second law of thermodynamics asserts that energy has quality as well as quantity. By quality, it means the ability or work potential of a certain energy source having a certain amount of energy to cause change, *i.e.*, the amount of energy that can be extracted as useful work that is termed as exergy. First and second law efficiencies are often called energy and exergy efficiencies, respectively. It is expected that exergy efficiencies are usually lower than the energy efficiencies, because the irreversibilities of the process destroy some of the input exergy.[10]

Exergy analysis method is employed to detect and to evaluate quantitatively the causes of the thermodynamic imperfection of the process under consideration. It can, therefore, indicate the possibilities of thermodynamic improvement of the process under consideration, but only an economic analysis can decide the expediency of a possible improvement.[11]

12.2 Energetic and Exergetic Relations

12.2.1 Dead (or Reference) State

Exergy is always evaluated with respect to a reference environment (*i.e.,* dead state). When a system is in equilibrium with the environment, the state of the system is called the dead state due to the fact that the exergy is zero. At the dead state, the conditions of mechanical, thermal, and chemical equilibrium between the system and the environment are satisfied: the pressure, temperature, and chemical potentials of the system equal those of the environment, respectively. In addition, the system has no motion or elevation relative to coordinates in the environment. Under these conditions, there is neither the possibility of a spontaneous change within the system or the environment nor an interaction between them. The value of exergy is zero. Another type of equilibrium between the system and environment can be identified. This is a restricted form of equilibrium, where only the conditions of mechanical and thermal equilibrium (thermomechanical equilibrium) must be satisfied. Such a state is called the restricted dead state. At the restricted dead state, the fixed quantity of matter under consideration is imagined to be sealed in an envelope impervious to mass flow, at zero velocity and elevation relative to coordinates in the environment, and at the temperature T_0 and pressure P_0, often taken as 25 °C and 1 atm.[12]

12.2.2 Relations Used

For a general steady-state, steady-flow process, the four balance (mass, energy, entropy and exergy) equations are employed to find the work and heat interactions, the rate of exergy decrease, the rate of irreversibility, and the energy and exergy efficiencies.[8,13–19]

12.2.2.1 Mass- and Energy-Balance Equations

The mass-balance equation can be expressed in the rate form as

$$\sum \dot{m}_{in} = \sum \dot{m}_{out} \tag{12.1}$$

where $\dot{m}$ is the mass flow rate, and the subscript *in* stands for inlet and *out* for outlet.

The general energy balance can be expressed below as the total energy input equal to total energy output ($\dot{E}_{in} = \dot{E}_{out}$), with all energy terms as follows:

$$\dot{Q} + \sum \dot{m}_{in} h_{in} = \dot{W} + \sum \dot{m}_{out} h_{out} \tag{12.2}$$

where $\dot{Q} = \dot{Q}_{net,in} = \dot{Q}_{in} - \dot{Q}_{out}$ is the rate of net heat input, $\dot{W} = \dot{W}_{net,out} = \dot{W}_{out} - \dot{W}_{in}$ is the rate of net work output, and h is the specific enthalpy.

Assuming no changes in kinetic and potential energies with no heat or work transfers, the energy balance given in eqn (12.2) can be simplified to flow enthalpies only:

$$\sum \dot{m}_{in} h_{in} = \sum \dot{m}_{out} h_{out} \tag{12.3}$$

12.2.2.2 Entropy- and Exergy-Balance Equations

The rate form of the entropy balance can be expressed as

$$\dot{S}_{in} - \dot{S}_{out} + \dot{S}_{gen} = 0 \tag{12.4}$$

where the rates of entropy transfer by heat transferred at a rate of $\dot{Q}_k$ through the boundary at temperature T_k at location k and mass flowing at a rate of $\dot{m}$ are $\dot{S}_{heat} = \dot{Q}_k / T_k$ and $\dot{S}_{mass} = \dot{m}s$, respectively.

Taking the positive direction of heat transfer to be to the system, the rate form of the general entropy relation given in eqn (12.4) can be rearranged to give

$$\dot{S}_{gen} = \sum \dot{m}_{out} s_{out} - \sum \dot{m}_{in} s_{in} - \sum \frac{\dot{Q}_k}{T_k} \tag{12.5}$$

It is also usually more convenient to find $\dot{S}_{gen}$ first and then to evaluate the exergy destroyed or the irreversibility rate $\dot{I}$ directly from the following equation,

$$\dot{I} = \dot{E}x_{dest} = T_0 \dot{S}_{gen} \tag{12.6}$$

The total exergy of a system $\dot{E}x$ can be divided into four components, namely (i) physical exergy $\dot{E}x^{ph}$, (ii) kinetic exergy $\dot{E}x^{kn}$, (iii) potential exergy $\dot{E}x^{pt}$, and

(iv) chemical exergy $\dot{E}x^{ch}$,[5]

$$\dot{E}x = \dot{E}x^{ph} + \dot{E}x^{kn} + \dot{E}x^{pt} + \dot{E}x^{ch} \tag{12.7}$$

Although exergy is an extensive property, it is often convenient to work with it on a unit of mass or molar basis. The total specific exergy on a mass basis may be written as follows:

$$ex = ex^{ph} + ex^{kn} + ex^{pt} + ex^{ch} \tag{12.8}$$

Physical exergy is the majority of the GSHP system. Therefore, chemical exergy, potential exergy, nuclear exergy, magnetic exergy, and kinetic exergy (kinetic energy) were neglected in this study.

The general exergy balance can be expressed in the rate form as

$$\underbrace{\dot{E}x_{in} - \dot{E}x_{out}}_{\substack{\text{Rate of net exergy transfer} \\ \text{by heat, work, and mass}}} = \underbrace{\dot{E}x_{dest}}_{\text{Rate of exergy destruction}} \tag{12.9a}$$

or

$$\dot{E}x_{heat} - \dot{E}x_{work} + \dot{E}x_{mass,in} - \dot{E}x_{mass,out} = \dot{E}x_{dest} \tag{12.9b}$$

Using eqn (12.9b), the rate form of the general exergy balance can also be written as

$$\sum\left(1 - \frac{T_0}{T_k}\right)\dot{Q}_k - \dot{W} + \sum \dot{m}_{in}\psi_{in} - \sum \dot{m}_{out}\psi_{out} = \dot{E}x_{dest} \tag{12.10}$$

with

$$\psi = (h - h_0) - T_0(s - s_0) \tag{12.11a}$$

where ψ is the flow (specific) exergy and the subscript zero indicates properties at the restricted dead state of P_0 and T_0.

Multiplying flow or specific exergy given in eqn (12.11a) by the mass flow rate of the fluid (refrigerant or water) gives the exergy rate

$$\dot{E}x = \dot{m}[(h - h_0) - T_0(s - s_0)] \tag{12.11b}$$

The specific exergy (flow exergy) of an incompressible substance (*i.e.*, water) is given by[20]

$$\psi_w = C\left(T - T_0 - T_0 \ln\frac{T}{T_0}\right) \tag{12.12a}$$

The total flow exergy of air is calculated from[21]

$$\psi_{a,t} = (C_{p,a} + \omega C_{p,v})T_0[(T/T_0) - 1 - \ln(T/T_0)] + (1 + 1.6078\omega)R_a T_0 \ln(P/P_0) + R_a T_0 \left\{ \begin{array}{l} (1 + 1.6078\omega)\ln[(1 + 1.6078\omega_0)/ \\ (1 + 1.6078\omega)] + 1.6078\omega \ln(\omega/\omega_0) \end{array} \right\} \tag{12.12b}$$

where the specific humidity ratio is

$$\omega = \dot{m}_v/\dot{m}_a \tag{12.13}$$

Assuming air to be a perfect gas, the specific physical exergy of air is calculated by the following relation[18]

$$\psi_{a,per} = C_{p,a}\left(T - T_0 - T_0 \ln\frac{T}{T_0}\right) + R_a T_0 \ln\frac{P}{P_0} \tag{12.14}$$

12.2.2.3 Energy and Exergy Efficiencies

Basically, the energy efficiency of any system can be defined as the ratio of total energy output rate to total energy input rate

$$\eta = \frac{\dot{E}_{output}}{\dot{E}_{input}} \tag{12.15}$$

where in most cases "*output*" refers to "*useful*" one.

However, there are numerous ways of formulating exergetic (or exergy or second-law) efficiency (effectiveness, or rational efficiency) for various energy systems. It is very useful to define efficiencies based on exergy (sometimes called *second law efficiencies*). Whereas there is no standard set of definitions in the literature, two different approaches are generally used—one is called "*universal* (or *brute-force)*", while the other is called "*functional*".[22]

The universal exergy efficiency for any system is defined as the ratio of the sum of all output exergy terms to the sum of all input exergy terms. The functional exergy efficiency for any system is defined as the ratio of the exergy associated with the desired energy output to the exergy associated with the energy expended to achieve the desired output.

Here, in a similar way, exergy efficiency is defined as the ratio of total exergy output to total exergy input, *i.e.*,

$$\varepsilon = \frac{\dot{E}x_{output}}{\dot{E}x_{input}} = 1 - \frac{\dot{E}x_{dest}}{\dot{E}x_{input}} \tag{12.16}$$

where "output or out" stands for "net output" or "product" or "desired value" or "benefit", and "input or in" stands for "given" or "used" or "fuel".

It is clear that the universal definition can be applied in a straightforward manner, irrespective of the nature of the component, once all exergy flows have been determined. The functional definition, however, requires judgment and a clear understanding of the purpose of the system under consideration before the working equation for the efficiency can be formulated.[22]

The exergy efficiency of a heat exchanger is determined by the increase in the exergy of the cold stream divided by the decrease in the exergy of the hot stream on a rate basis as follows:

$$\psi_{HE} = \frac{\dot{E}x_{cold,out} - \dot{E}x_{cold,in}}{\dot{E}x_{hot,in} - \dot{E}x_{hot,out}} = \frac{\dot{m}_{cold}(ex_{cold,out} - ex_{cold,in})}{\dot{m}_{hot}(ex_{hot,in} - ex_{hot,out})} \tag{12.17}$$

12.2.2.4 Exergetic Improvement Potential and Some Thermodynamic Parameters

Van Gool[23] has also proposed that maximum improvement in the exergy efficiency for a process or system is obviously achieved when the exergy loss or irreversibility ($\dot{E}x_{in}$–$\dot{E}x_{out}$) is minimized. Consequently, he suggested that it is useful to employ the concept of an exergetic *"improvement potential"* when analyzing different processes or sectors of the economy. This improvement potential in the rate form, denoted $E\dot{P}$, is given by

$$I\dot{P} = (1 - \varepsilon)(\dot{E}x_{in} - \dot{E}x_{out}) \tag{12.18}$$

In the thermodynamics analysis, the following parameters may also be used:[24]

Fuel depletion ratio:

$$\delta_i = \dot{I}_i / \dot{F}_T \tag{12.19}$$

Relative exergy destruction (irreversibility):

$$\chi_i = \dot{I}_i / \dot{I}_T \tag{12.20}$$

Productivity lack:

$$\xi_i = \dot{I}_i / \dot{P}_T \tag{10.21}$$

Exergetic factor:

$$f_i = \dot{F}_i / \dot{F}_T \tag{12.22}$$

12.3 Application of Exergy Analysis to Various Compound Energy Systems

In the following, the exergy analysis method presented here is applied to two various systems. The first one[8] is related to a solar-assisted DHW tank integrated ground-source heat pump system, while the second one[1] includes a high-temperature electrolysis (HTE) process coupled with a geothermal source.

12.3.1 Case Study 1

12.3.1.1 System Description

Figure 12.1 shows a schematic of the solar-assisted DHW tank integrated GSHP system used in a 180 m^2 private residence, to which the model given here is applied in the heating mode since the experimental data were available in this mode only. This system was designed for heating/cooling the residence and DHW needs. The detailed description of the system is given elsewhere.[8,26] Solar heat is used in priority to heat DHW and is injected into the ground via boreholes only when the DHW temperature setting is reached.

As can be seen in Figure 12.1, the system consists of mainly five parts, namely (i) a water-to-water heat pump unit with a heating capacity of 15.8 kW, (ii) a ground heat exchanger system consisting of two U-boreholes with an individual depth of 90 m, (iii) a solar collector system consisting of rooftop thermal solar collectors covering 12 m^2, (iv) a DHW tank with a electrical supplementary heater, (v) a heating/cooling floor with a total surface of 154 m^2, and (vi) circulating pumps.

12.3.1.2 System Analysis

Mass and energy balances as well as exergy destructions obtained from exergy balances for each of the solar-assisted DHW tank integrated GSHP system components illustrated in Figure 12.1 are derived as follows:

Compressor (I):

$$\dot{m}_1 = \dot{m}_{2,s} = \dot{m}_{act,s} = \dot{m}_r \tag{12.23a}$$

$$\dot{W}_{comp} = \dot{m}_r(h_{2,act} - h_1) \tag{12.23b}$$

$$\dot{E}x_{dest,comp} = \dot{m}_r(\psi_1 - \psi_{2,act}) + \dot{W}_{comp,elec} \tag{12.23c}$$

$$\dot{W}_{comp,elec} = \dot{W}_{comp}/(\eta_{comp,elec}\eta_{comp,mech}) \tag{12.23d}$$

$$\dot{W}_{com,elec} = \sqrt{3}\, V_{comp} I_{comp} \cos\,\varphi \tag{12.23e}$$

$$\dot{E}x_{dest,comp,mech,elec} = \dot{W}_{comp,elec}(1 - \eta_{comp,elec}\eta_{comp,mech}) \tag{12.23f}$$

$$\dot{E}x_{dest,comp,\text{int}} = \dot{E}x_{dest,comp} - \dot{E}x_{dest,comp,mech,elec} \tag{12.23g}$$

where heat interactions with the environment are neglected.

Condenser (II):

$$\dot{m}_2 = \dot{m}_3 = \dot{m}_r; \quad \dot{m}_{14} = \dot{m}_{17} = \dot{m}_{w,fh} \tag{12.24a}$$

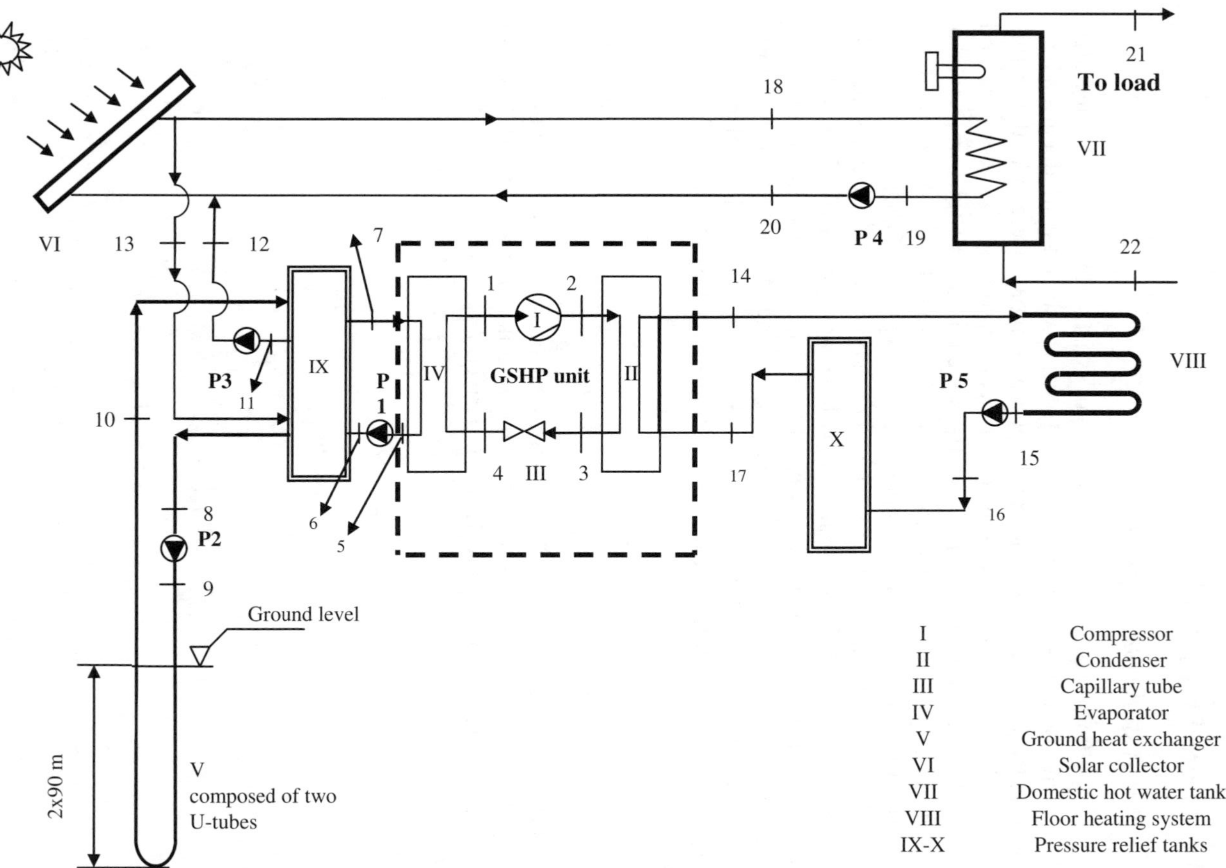

Figure 12.1 A schematic of the solar-assisted domestic hot water tank integrated GSHP system in the case study 1.[8]

$$\dot{Q}_{cond} = \dot{m}_r(h_{2,act} - h_3); \quad \dot{Q}_{cond} = \dot{m}_{w,fh}C_{p,w,fh}(T_{14} - T_{17}) \tag{12.24b}$$

$$\dot{E}x_{dest,cond} = \dot{m}_r(\psi_{2,act} - \psi_3) + \dot{m}_{w,fh}(\psi_{14} - \psi_{17}) \tag{12.24c}$$

Expansion (throttling) valve (III):

$$\dot{m}_3 = \dot{m}_4 = \dot{m}_r \tag{12.25a}$$

$$h_3 = h_4 \tag{12.25b}$$

$$\dot{E}x_{dest,\exp} = \dot{m}_r(\psi_3 - \psi_4) \tag{12.25c}$$

Evaporator (IV):

$$\dot{m}_4 = \dot{m}_1 = \dot{m}_r; \quad \dot{m}_5 = \dot{m}_6 = \dot{m}_7 = \dot{m}_{w,prt9} \tag{12.26a}$$

$$\dot{Q}_{evap} = \dot{m}_r(h_1 - h_4); \quad \dot{Q}_{evap} = \dot{m}_{w,prt9}C_{w,prt9}(T_7 - T_5) \tag{12.26b}$$

$$\dot{E}x_{dest,evap} = \dot{m}_r(\psi_4 - \psi_1) + \dot{m}_{w,prt9}(\psi_7 - \psi_5) \tag{12.26c}$$

Ground heat exchanger (V):

$$\dot{m}_8 = \dot{m}_9 = \dot{m}_{10} = \dot{m}_{w,ghe} \tag{12.27a}$$

$$\dot{Q}_{ghe} = \dot{m}_{w,ghe}(h_{10} - h_8) \tag{12.27b}$$

$$\dot{E}x_{dest,ghe} = \dot{m}_w(\psi_9 - \psi_{10}) + \dot{Q}_{ghe}\left(1 - \frac{T_0}{T_{ghe}}\right) \tag{12.27c}$$

Solar collector (VI):

The useful exergy delivered and the exergy absorbed by the solar collector are given as follows, respectively.[8]

$$\dot{E}x_{dest,scol} = \dot{E}x_u + \dot{E}x_{scol} \tag{12.28a}$$

with

$$\dot{E}x_u = \dot{m}_{w,scol}[(h_{w,18} - h_{w,20}) - T_0(s_{w,18} - s_{w,20})] \tag{12.28b}$$

or

$$\dot{E}x_u = \dot{m}_{w,scol}C_{w,scol}\left[(T_{w,18} - T_{w,20}) - T_0\left(\ln\frac{T_{w,18}}{T_{w,20}}\right)\right] \tag{12.28c}$$

and

$$\dot{E}x_{scol} = AI_T\left[1 + \frac{1}{3}\left(\frac{T_0}{T_{sr}}\right)^4 - \frac{4}{3}\left(\frac{T_0}{T_{sr}}\right)\right] \tag{12.28d}$$

where T_{sr} is the solar radiation temperature and taken to be 6000 K, while the Petela expression is used in calculating the exergy of solar radiation as the exergy input to the solar collector.[26]

Domestic hot water tank (VII):

$$\dot{m}_{18} = \dot{m}_{19} = \dot{m}_{w,dhwt}; \quad \dot{m}_{21} = \dot{m}_{22} = \dot{m}_{w,load} \tag{12.29a}$$

$$\dot{E}x_{dest,dhwt} = (\dot{E}x_{18} + \dot{E}x_{22}) - (\dot{E}x_{19} + \dot{E}x_{21}) \tag{12.29b}$$

where heat interactions with the environment are neglected.

Floor heating system (VIII):

$$\dot{m}_{14} = \dot{m}_{15} = \dot{m}_{w,fhs} \tag{12.30a}$$

$$\dot{Q}_{fhs} = \dot{m}_{w,fhs}(h_{14} - h_{15}) \tag{12.30b}$$

$$\dot{E}x_{dest,fhs} = \dot{m}_{w,fhs}(\psi_{14} - \psi_{15}) - \dot{Q}_{fhs}\left(1 - \frac{T_0}{T_{fhs}}\right) \tag{12.30c}$$

Pressure relief tanks (IX–X):

$$\dot{E}x_{dest,prt} = \Sigma\dot{E}x_{in,prt} - \Sigma\dot{E}x_{out,prt} \tag{12.31}$$

Circulating pumps (P 1–P 4):

$$\dot{m}_5 = \dot{m}_{6s} = \dot{m}_{6,act} = \dot{m}_{w,prt9} \tag{12.32a}$$

$$\dot{W}_{pump} = \dot{m}_{w,prt9}(h_{6,act} - h_5) \tag{12.32b}$$

$$\dot{E}x_{dest,pump} = \dot{m}_{w,prt9}(\psi_5 - \psi_6) + \dot{W}_{pump,elec} \tag{12.32c}$$

$$\dot{W}_{pump,elec} = \dot{W}_{pump}/(\eta_{pump,elec}\eta_{pump,mech}) \tag{12.32d}$$

$$\dot{W}_{pump,elec} = V_{pump}I_{pump}\cos\varphi \tag{12.32e}$$

where the relations for the circulating pump (P 1) only are given here.

The exergy efficiencies of the GSHP drying system components and the whole system are evaluated as follows:

Compressor (I):

$$\varepsilon_{comp} = \frac{\dot{E}x_{2,act} - \dot{E}x_1}{\dot{W}_{comp,elec}} \tag{12.33}$$

Condenser (II):

$$\varepsilon_{cond} = \frac{\dot{E}x_{14} - \dot{E}x_{17}}{\dot{E}x_{2,act} - \dot{E}x_3} = \frac{\dot{m}_{w,fh}(\psi_{14} - \psi_{17})}{\dot{m}_r(\psi_{2,act} - \psi_3)} \tag{12.34}$$

Expansion (throttling) valve (III):

$$\varepsilon_{\exp} = \frac{\dot{E}x_4}{\dot{E}x_3} = \frac{\psi_4}{\psi_3} \tag{12.35}$$

Evaporator (IV):

$$\varepsilon_{evap} = \frac{\dot{E}x_5 - \dot{E}x_7}{\dot{E}x_4 - \dot{E}x_1} = \frac{\dot{m}_{w,prt9}(\psi_5 - \psi_7)}{\dot{m}_r(\psi_4 - \psi_1)} \tag{12.36}$$

Ground heat exchanger (V):

$$\varepsilon_{ghe} = \frac{\dot{E}x_{10}}{\dot{E}x_8 + \dot{Q}_{ghe}[1 - (T_0/T_{ghe})]} = \frac{\dot{m}_{w,ghe}\,\psi_{10}}{\dot{m}_{w,ghe}\,\psi_8 + \dot{Q}_{ghe}[1 - (T_0/T_{ghe})]} \tag{12.37}$$

Solar collector (VI):

$$\varepsilon_{scol} = \frac{\dot{E}x_u}{\dot{E}x_{scol}} \tag{12.38}$$

Domestic hot water tank (VII):

$$\varepsilon_{dhwt} = \frac{\dot{E}x_{21} - \dot{E}x_{22}}{\dot{E}x_{18} - \dot{E}x_{19}} = \frac{\dot{m}_{w,load}(\psi_{21} - \psi_{22})}{\dot{m}_{w,dhwt}(\psi_{18} - \psi_{20})} \tag{12.39}$$

Floor heating system (VIII):

$$\varepsilon_{fhs} = \frac{\dot{Q}_{fhs}\left(1 - \frac{T_0}{T_{fhs}}\right)}{\dot{E}x_{14} - \dot{E}x_{15}} = \frac{\dot{Q}_{fhs}\left(1 - \frac{T_0}{T_{fhs}}\right)}{\dot{m}_{w,fhs}(\psi_{14} - \psi_{15})} \tag{12.40}$$

Circulating pumps (P 1–P 5):

$$\varepsilon_{pump} = \frac{\dot{E}x_6 - \dot{E}x_5}{\dot{W}_{pump,elec}} = \frac{\dot{m}_{w,prt9}(\psi_6 - \psi_5)}{\dot{W}_{pump,elec}} \tag{12.41}$$

which is given for the circulating pump (P 1) only.

GSHP unit (I-IV):

$$\varepsilon_{GSHP} = \frac{\dot{E}x_{heat}}{\dot{W}_{comp,elec}} = \frac{\dot{E}x_{14} - \dot{E}x_{17}}{\dot{W}_{comp,elec}} \tag{12.42}$$

Overall GSHP system:

$$\varepsilon_{GSHP,sys} = \frac{\dot{E}x_{14} - \dot{E}x_{17}}{\dot{W}_{comp,elec} + \Sigma \dot{W}_{pump,elec}} \tag{12.43}$$

The exergetic coefficients of performance of the GSHP unit and whole system are as follows:

$$COP_{ex,GSHP} = \frac{\dot{Q}_{cond}\left(1 - \frac{T_0}{T_{cond}}\right)}{\dot{W}_{comp,elec}} \tag{12.44a}$$

$$COP_{ex,sys} = \frac{\dot{Q}_{cond}\left(1 - \frac{T_0}{T_{cond}}\right)}{\dot{W}_{comp,elec} + \Sigma \dot{W}_{pump,elec}} \tag{12.44b}$$

12.3.1.3 Results and Discussion

The system given in Figure 12.1 is evaluated from the exergetic point of view using the analysis method presented in Section 12.3.1.2.

The following assumptions are made for the exergy analysis of the system given as an illustrative example.

(a) All processes are steady state and steady flow with negligible potential and kinetic energy effects and no chemical or nuclear reactions.
(b) The directions of heat transfer to the system and work transfer from the system are positive.
(c) The pressure losses in the pipelines connecting the components are ignored, since their lengths are short.
(d) The compressor mechanical ($\eta_{comp,mech}$) and the compressor motor electrical ($\eta_{comp,elec}$) efficiencies are assumed to be 83% and 90%, respectively.
(e) The adiabatic efficiency of the compressor is taken to be 80%.
(f) The circulating pump mechanical ($\eta_{pump,mech}$) and the circulating pump motor electrical ($\eta_{pump,elec}$) efficiencies are 82 and 88%, respectively.
(g) The heat extraction rate from the ground is 39.7 W/m per borehole length, while the evaporator capacity is found to be 7.14 kW for a

borehole length of 180 m, while the ground temperature is taken to be 15 °C.

(h) The fluid circulating in the solar collector and ground heat exchanger circuits consists of a 35% propylene glycol solution with a constant specific heat value of 3.7 kJ/kg K.

(i) Electrical consumption of circulating pumps (P 1, P 2 and P 5) amounts to 14% of the GSHP unit's electrical consumption.

(j) The refrigerant used in the GSHP unit is R-410a.

(k) The experimental data used in the calculations are taken from ref. 25, while some calculations are also made using the values given in this reference in order to complete Table 12.1 where the state numbers for one representative unit specified in Figure 12.1 are included. The thermodynamic properties of water and the refrigerant are found using Engineering Equation Solver (shortly the so-called ESE) software package program. Note that here state 0 indicates the reference state for both fluids.

(l) The floor surface temperature is assumed to be 27 °C.

(m) The total solar radiation on the horizontal surface is taken to be on 700 W/m^2 and the total collector area is 12 m^2.

(n) The values for the dead (reference) state temperature and pressure are taken to be 19 °C and 101.325 kPa, respectively.

Temperature, pressure and mass flow rate data for the refrigerant R-410a, the water and water–antifreeze solution are shown in Table 12.1 following the state numbers specified in Figure 12.1. This table also includes exergy rates calculated for each state.

Figure 12.2 indicates the exergy destruction (irreversibility) rates of the system components at a reference state temperature of 19 °C. The greatest exergy destruction rate on the GSHP unit basis occurs in the condenser, followed by the compressor, the expansion valve and the evaporator.

The mechanical-electrical losses are calculated using eqns (12.23f) and (12.23g) and are found to account for 25.22% of the system input. The mechanical-electrical losses are due to imperfect electrical, mechanical and isentropic efficiencies and emphasize the need for paying close attention to the selection of this equipment, since components of inferior performance can considerably reduce overall system performance. Since compressor power depends strongly on the inlet and outlet pressures, any heat-exchanger improvements that reduce the temperature difference will reduce compressor power by bringing the condensing and evaporating temperatures closer together. From a design standpoint, the compressor destruction rate can be reduced independently. Recent advances in the market have led to the use of scroll compressors. The destruction rates in the evaporator and the condenser occur due to the temperature differences between the two heat exchanger fluids, pressure losses, flow imbalances and heat transfer with the environment. The destruction rate associated with the expansion valve (capillary tube) is due to the pressure drop of the refrigerant passing through it. The only way to eliminate

Table 12.1 Energy and exergy analyses results of the case study 1.[8]

State No.	*Description*	*Fluid*	*Phase*	*Temperature, T (°C)*	*Pressure, P (kPa)*	*Specific enthalpy, h (kJ/kg)*	*Specific Entropy, s (kJ/kg K)*	*Mass flow rate, $\dot{m}$ (kg/s)*	*Specific exergy, ψ (kJ/kg)*	*Energy rate, $\dot{E}=\dot{m}h$ (kW)*	*Exergy rate, $\dot{E}x=\dot{m}\psi$ (kW)*
0	-	Refrigerant (R-410a)	Dead state	19	101.325	316.1	1.394	-	-	-	-
0″	-	Water	Dead state	19	101.325	79.7	0.282	-	-	-	-
0″	-	Water–anti-freeze mixture	Dead state	19	101.325	70.3	0.249	-	-	-	-
1	Evaporator outlet/ Compressor inlet	Refrigerant	Superheated vapor	–2	677.1	283.1	1.070	0.051	61.66	14.44	3.14
2,s	Condenser inlet/ Compressor outlet	Refrigerant	Superheated vapor	68.6	2724.1	322.2	1.070	0.051	100.76	16.43	5.14
2,act	Condenser inlet/ Compressor outlet	Refrigerant	Superheated vapor	75.9	2724.1	331.9	1.098	0.051	102.28	16.93	5.22
3	Condenser outlet/ Expansion valve inlet	Refrigerant	Saturated liquid	45	2724.1	135.6	0.488	0.051	84.19	6.92	4.29
4	Expansion valve outlet/ Evaporator inlet	Refrigerant	Mixture	–5	677.1	135.6	0.520	0.051	74.84	6.92	3.82
5	Circulating pump inlet (P 4)	Water–anti-freeze mixture	Compressed liquid	2		7.4	0.027	0.508	1.96	3.76	0.99
6	Circulating pump outlet (P 4)/Pressure relief tank inlet	Water–anti-freeze mixture	Compressed liquid	2.1		7.77	0.028	0.508	2.04	3.95	1.04
7	Pressure relief tank outlet (Evaporator side)	Water–anti-freeze mixture	Compressed liquid	6		22.2	0.080	0.508	1.27	11.28	0.65
8	Circulating pump inlet (P 2)	Water–anti-freeze mixture	Compressed liquid	2.1		7.77	0.028	0.508	2.04	3.95	1.04

9	Circulating pump outlet (P 2)/ Ground heat exchanger inlet	Water–anti-freeze mixture	Compressed liquid	2.2	8.14	0.030	0.508	1.82	4.14	0.92
10	Ground heat exchanger outlet	Water–anti-freeze mixture	Compressed liquid	6	22.2	0.080	0.508	1.27	11.28	0.65
14	Floor-heating system supply	Water	Compressed liquid	35	146.7	0.505	0.345	1.85	50.61	0.64
15	Floor-heating system return/Circulating pump inlet (P 5)	Water	Compressed liquid	28	117.4	0.409	0.345	0.60	40.50	0.21
16	Circulating pump outlet (P 5)/Pressure-relief tank inlet	Water	Compressed liquid	28.1	117.8	0.410	0.345	0.71	40.64	0.24
17	Pressure-relief tank outlet/Condenser inlet (water side)	Water	Compressed liquid	28.1	117.8	0.410	0.345	0.71	40.64	0.24
18	Solar collector outlet/Domestic hot water tank inlet	Water–anti-freeze mixture	Compressed liquid	71	297.3	0.967	0.350	17.24	104.06	6.03
19	Domestic hot water tank outlet/Circulating pump inlet (P 4)	Water–anti-freeze mixture	Compressed liquid	67	280.5	0.918	0.350	14.75	98.18	5.16
20	Circulating pump outlet (P 4)/Solar collector inlet	Water–anti-freeze mixture	Compressed liquid	67.1	280.9	0.919	0.350	14.86	98.32	5.20
21	Hot-water supply (to load)	Water	Compressed liquid	65	272.1	0.893	0.034	13.90	9.25	0.47
22	Cold-water inlet	Water	Compressed liquid	30	125.8	0.436	0.034	1.11	4.28	0.04

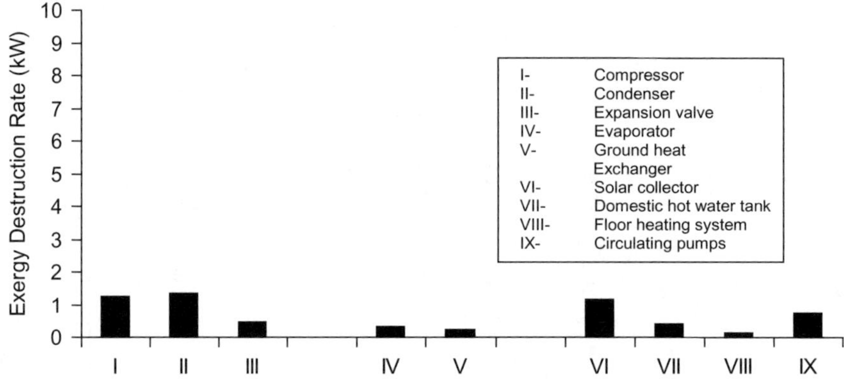

Figure 12.2 Exergy destruction (irreversibility) rates of the system components at a reference state temperature of 19 °C for the case study 1.

the throttling loss is to replace the capillary tube with an isentropic turbine (an isentropic expander) and to recover some shaft work from the pressure drop.

The greatest exergy destruction rates on the whole system basis occurs in the solar collector with 8.68 kW, followed by the components, such as the condenser with 1.33 kW, the compressor with 1.25 kW, the expansion valve with 0.47 and the DHW tank with 0.44 kW, as shown in Figure 12.2.

Figure 12.3 illustrates the variation of exergy-efficiency values for the GSHP system components. It is obvious from this figure that the expansion valve has the highest value, followed by the ground source heat exchanger, the floor heating system, the compressor, the evaporator, the DHW tank, the condenser and the solar collector, varying between about 11 and 89%.

The exergy efficiency of the solar collector is calculated to be relatively low compared to the other system components, while its exergy destruction has the highest value among the other components of the entire system. Solar energy is highly inefficient from the standpoint of avoiding the one-way destruction of exergy. Therefore, the sun should be first used in high-temperature applications; the heat rejected from these first applications can be then cascaded to applications at lower temperatures, eventually, to the task of keeping a building warm, as denoted by Bejan.[27]

The exergy efficiencies of the whole system are calculated in two ways, namely using eqns (12.42) and (12.43) based on the desired effect (benefit)/fuel basis, respectively and on the product/fuel basis using the values given in Table 12.1. They are found to be 12.01 and 9.88% based on the benefit/fuel basis, respectively, while the exergy-efficiency value for the GSHP unit is obtained to be 72.33% on the product/fuel basis. This means that the efficiency values differ from each other depending on the exergetic efficiency definition used.

Van Gool's exergetic improvement potential rates ($\dot{IP}$) are calculated using eqn (12.18) for each component of the GSHP unit, while they are shown in

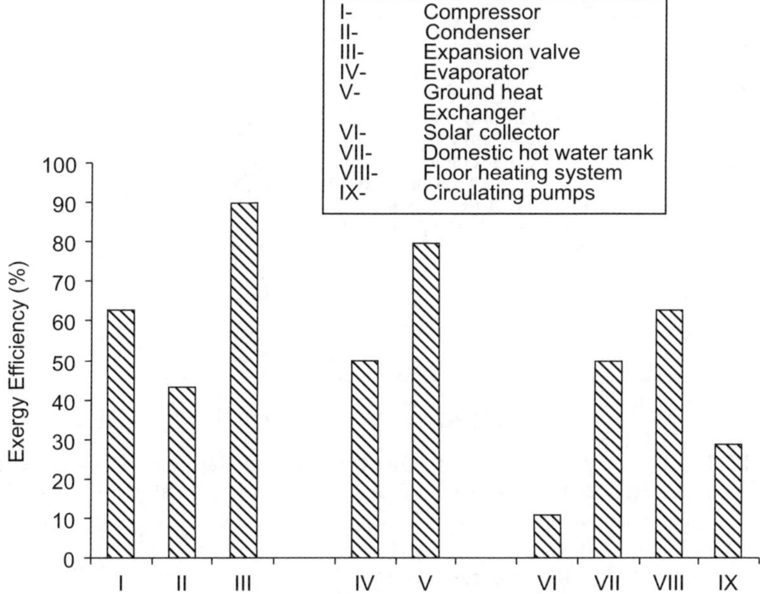

Figure 12.3 Exergy-efficiency values of the system components at a reference state temperature of 19 °C for the case study 1.

Figure 12.4. It is found that the condenser has the highest $I\dot{P}$ value, followed by the compressor, the evaporator and the expansion valve.

Using eqn (12.20), the relative exergy destructions values are obtained to be 39.23% for the condenser, 36.87% for the compressor, 13.86% for the expansion valve and 10.04% for the evaporator on the GSHP unit basis.

12.3.2 Case Study 2

In the literature, many studies have recently been conducted on future hydrogen demand; for example, one has been presented by Sigurvinsson *et al.*,[28] considering a HTE process coupled with a geothermal source.

12.3.2.1 System Description

The schematic flow diagram of the HTE process along with its main components is shown in Figure 12.5. As can be seen in this figure, geothermal steam enters the heat exchanger system of HTE system at #1 and #5. The entering temperature of the electrolyzer could be between 750 and 900 °C. To be effective from a thermodynamic point of view, the HTE requires a recovery of the heat contained at the exit from the electrolyzer. Heat needs to be recovered in parallel from oxygen and from the hydrogen-steam mixture in order to heat the steam in contact with the geothermal source, up to the desired temperature at

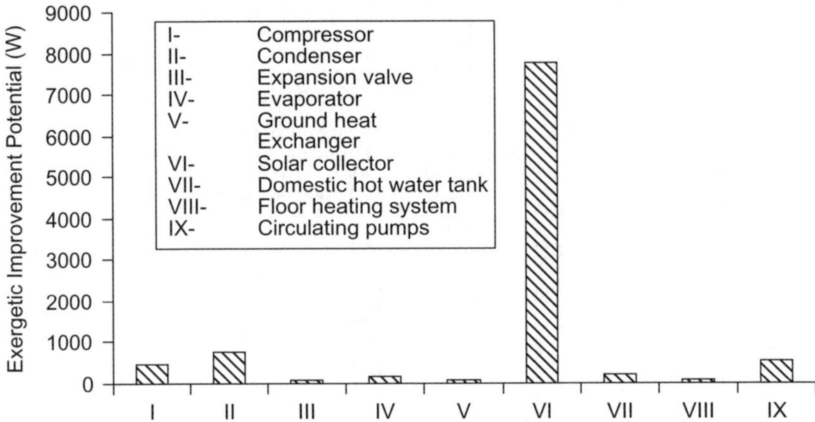

Figure 12.4 Exergetic improvement potential rates of the system components at a reference state temperature of 19 °C for the case study 1.

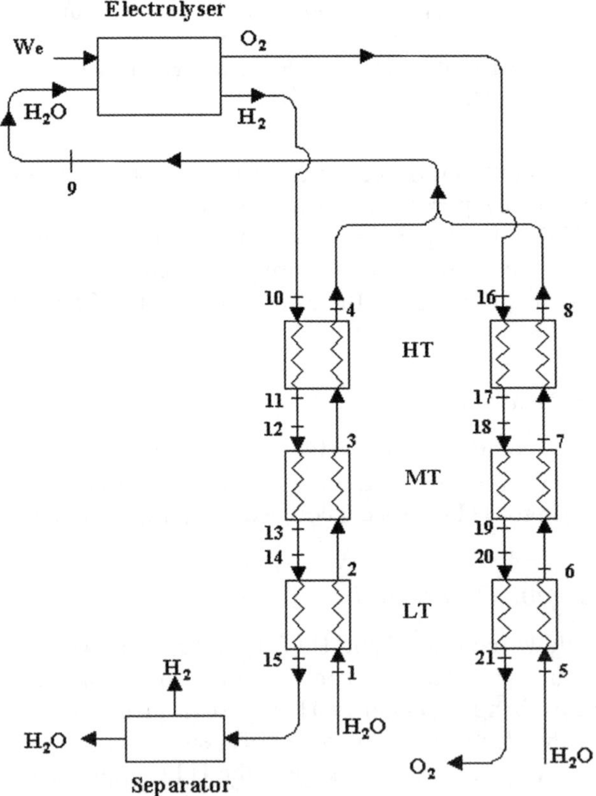

Figure 12.5 A schematic of the HTE system in the case study 2, as proposed by Sigurvinsson *et al.*[28]

the entry of the electrolyzer. There is also a separator in this system, which separates hydrogen from the hydrogen-steam mixture. As for the heat-exchanger networks, counter current heat exchangers are used, while the inlet temperature of the electrolyzers is kept as 950 °C. The temperatures in the geothermal case vary between 200 and 950 °C. The heat exchangers are classified into three groups according to the ranges of the temperatures since this temperature range cannot be covered with one type of heat exchangers. Therefore, (i) Low temperature (LT): stainless heat exchanger, $T<600\,^{\circ}C$ and 7 MPa, (ii) Medium temperature (MT): nickel-based heat exchanger, $600\,^{\circ}C<T<850\,^{\circ}C$ and 7 MPa, and (iii) High temperature (HT): ceramic-based heat exchanger, $T>850\,^{\circ}C$ and 10–50 MPa.[28,29]

12.3.2.2 System Analysis

In this section, energy and exergy analyses of a high-temperature electrolysis are performed for comparison purposes. Before getting into energy and exergy analyses, the following assumptions are made:

- The values for the reference environment (dead state) temperature (T_0) and pressure (P_0) are 25 °C and 100 kPa, respectively.
- All processes are considered steady-state and steady-flow with negligible potential and kinetic energy effects in an adiabatic form.

For a general steady-state, steady-flow process, the mass, energy and exergy balance relations given by eqns (12.1), (12.2) and (12.9a), respectively, are employed to find the work input, the rate of exergy destruction, energy and energy efficiencies. Since mass is conserved in chemical reactions, the masses of products and reactants are equal and in general, the mass-balance equation can be expressed in the rate form as

$$\sum \dot{m}_R = \sum \dot{m}_P \tag{12.45}$$

where $\dot{m}$ is the mass flow rate, and the subscripts R and P stand for reactant and reactant, respectively.

The general energy balance is written as

$$\dot{E}_{in} - \dot{E}_{out} = \Delta \dot{E}_s \tag{12.46}$$

which becomes

$$Q + W = \sum n_P(\bar{h}_f^o + \bar{h} + \bar{h}^o)_P \sum n_R(\bar{h}_f^o + \bar{h} + \bar{h}^o)_R \tag{12.47}$$

and the exergy balance for the process, involving chemical reactions, becomes

$$\sum \dot{E}x_{in} - \sum \dot{E}x_{out} - \dot{E}x_{dest} = \Delta \dot{E}x_{sys} \tag{12.48}$$

For a steady-state system, $\Delta \dot{E}_{sys}$ is zero. The exergy associated with a process at a specified state is the sum of two contributions, namely physical and chemical.

Thus, the specific exergy of the process is calculated by

$$ex = (h - h_0) - T_0(s - s_0) + ex^{ch} \tag{12.49}$$

Combining eqns (12.48) and (12.49) yields,

$$ex_{dest} = \sum \left[(h - h_0) - T_0(s - s_0) + ex^{ch}\right] - \sum \left[(h - h_0) - T_0(s - s_0) + ex^{ch}\right]_{out} + W_{elec,in} \tag{12.50}$$

After writing mass, energy and exergy balances for the system, the enthalpy values of H_2, O_2 and H_2O are evaluated with Shomate equations given by

$$\bar{h} - \bar{h}_0 = AT + B(T^2/2) + C(T^3/3) + D(T^4/4) - \mathrm{E/T + F - H} \tag{12.51}$$

where T is 1/1000 of the specified temperature (in K) of compound and A, B, C, D, E, F, G and H are constants, as given in Table 12.2 for H_2, O_2 and H_2O.[30,31]

The chemical exergy based on a typical exergy reference environment exhibiting standard values of the environmental temperature T_0 and pressure P_0, such as 298 K and 100 kPa, is called standard chemical exergy. The values of the chemical exergies for the reactants and products are given in Table 12.3.

Table 12.2 Enthalpy of formation and Shomate constants along with Gibbs free energy and standard chemical exergy values for H_2, O_2 and H_2O. (Adapted from refs 30,31.).

		H_2O (g)	O_2 (g)	H_2 (g)
Enthalpy of formation and Shomate constants	$\bar{h}_f^0$(kJ/kmol)	–241830	0	0
	A	30.0920	29.6590	33.0661
	B	6.832514	6.137261	–11.36340
	C	6.793435	–1.186521	11.432816
	D	–2.534480	0.095780	–2.772874
	E	0.082139	–0.219663	–0.158558
	F	–250.881	–9.861391	–9.980797
	G	223.3967	237.9480	172.7079
	H	–241.8264	0	0
Specific Gibbs free energy of formation $\bar{g}_f^0$ (kJ/kmol)		–228638	0	0
Standard chemical exergy $e\bar{x}^{ch}$ (kJ/kmol)		9437	3970	236090

Table 12.3 Energy and exergy analysis results of the case study 2.[1] (*State numbers refer to Figure 12.5.).

State No	*Substance*	*Description*	*T(K)*	*P (kPa)*	$\dot{m}(kg/s)$	*h(kJ/kg)*	*s (kJ/kg.K)*	$ex^{ph}(kJ/kg)$	$\dot{E}x^{ph}(kW)$	$\dot{E}(kW)$
0	H_2O	Reference State	298	101.325	–	104.8	0.3669	–	–	–
0′	H_2	Reference State	298	101.325	–	0	64.82	–	–	–
0″	O_2	Reference State	298	101.325	–	0	6.407	–	–	–
1	H_2O	H_2O inlet LT-H_2	503	1500	11.400	2874	6.613	908	10350	32764
2	H_2O	H_2O outlet LT-H_2/H_2O inlet MT-H_2	753	1500	11.400	3429	7.513	1195	13619	39091
3	H_2O	H_2O outlet MT-H_2/H_2O inlet HT-H_2	978	1500	11.400	3932	8.096	1524	17373	44825
4	H_2O	H_2O outlet HT-H_2	1185	1500	11.400	4421	8.549	1878	21408	50399
5	H_2O	H_2O inlet LT-O_2	503	1500	4.000	2874	6.613	908	3631	11496
6	H_2O	H_2O outlet LT-O_2/H_2O inlet MT-O_2	693	1500	4.000	3299	7.332	1119	4474	13196
7	H_2O	H_2O outlet MT-O_2/H_2O inlet HT-O_2	983	1500	4.000	3944	8.108	1532	6129	15776
8	H_2O	H_2O outlet HT-O_2	1185	1500	4.000	4421	8.549	1878	7512	17684
9	H_2O	Electrolyser inlet	1185	1500	15.400	4421	8.549	1878	28920	68083
10	H_2	H2 inlet HT-H_2	1223	10000	1.147	13653	66.51	13149	15076	15653
10′	H_2O	H_2O inlet HT-H_2	1223	10000	5.082	4486	7.731	2187	11113	22798
11	H_2	H2 outlet HT-H_2	987	10000	1.147	10056	63.24	10527	12069	11529
11′	H_2O	H_2O outlet HT-H_2	987	10000	5.082	3903	7.202	1761	8951	19835
12	H_2	H2 inlet MT-H_2	987	2000	1.147	10056	69.88	8548	9800	11529
12′	H_2O	H_2O inlet MT-H_2	987	2000	5.082	3950	7.982	1576	8009	20074
13	H_2	H2 outlet MT-H_2	782	2000	1.147	7016	66.43	6536	7494	8044
13′	H_2O	H_2O outlet MT-H_2	782	2000	5.082	3488	7.457	1270	6456	17726
14	H_2	H2 inlet LT-H_2	782	7000	1.147	7016	61.27	8074	9257	8044
14′	H_2O	H_2O inlet LT-H_2	782	7000	5.082	3432	6.826	1402	7127	17441
15	H_2	H2 outlet LT-H_2	577	7000	1.147	4030	56.84	6408	7347	4620
15	H_2O	H_2O outlet LT-H_2	577	7000	5.082	2854	5.958	1083	5504	14504
16	O_2	O_2 inlet HT-O_2	1223	10000	9.200	955.3	6.634	888	8166	8789
17	O_2	O_2 outlet HT-O_2	987	10000	9.200	725.9	6.428	720	6621	6678
18	O_2	O_2 inlet MT-O_2	987	2000	9.200	725.9	6.816	604	5557	6678
19	O_2	O_2 outlet MT-O_2	782	2000	9.200	406.6	6.474	387	3557	3741
20	O_2	O_2 inlet LT-O_2	782	7000	9.200	406.6	6.149	483	4448	3741
21	O_2	O_2 outlet LT-O_2	569	7000	9.200	258.2	5.917	404	3719	2375

The exergy destructions in the heat exchangers and electrolyzers are calculated as

$$\dot{E}x_{dest,\ HE} = \dot{E}x_{in} - \dot{E}x_{out} \tag{12.52}$$

$$\dot{E}x_{dest,elec} = \dot{W}_{elec} - \left(\dot{E}x_{out} - \dot{E}x_{in}\right) \tag{12.53}$$

In a high-temperature steam electrolysis (HTSE), water is electrolyzed as follows:

$$H_2O + W_{elec} + Q_g \Rightarrow H_2 + (1/2)O_2 \tag{12.54}$$

The energy efficiency of the overall system can be defined as

$$\eta_{overall} = \frac{\dot{m}_{H_2} HHV_{H_2}}{\dot{W}_{elec,in} + \dot{Q}_{g,in}} \tag{12.55}$$

The exergy efficiency of the overall system is defined as the ratio of the exergy of the hydrogen (H_2), over the total exergy input, which can be determined from

$$\varepsilon_{overall} = \frac{\sum \dot{E}x_{H_2}}{\dot{W}_{elec,in} + \sum \dot{E}x_{in}} \tag{12.56}$$

The fluid in the first heat exchanger network is only oxygen. In the second heat exchanger, the fluid is a steam and hydrogen mixture. This mixture depends on the recycling ratio, which is taken to be 0.33 in these analyses.[1]

The mass flow rates of hydrogen and oxygen are calculated by the following relations, respectively:

$$\dot{m}_{out,H_2} = (1 - r)\dot{m}_{H_2O}(M_{H_2}/M_{H_2O}) + r\dot{m}_{H_2O} \tag{12.57}$$

$$\dot{m}_{out,O_2} = [(1 - r)\dot{m}_{H_2O}](M_{O_2}/M_{H_2O})/2 \tag{12.58}$$

12.3.2.3 Results and Discussion

The temperature, pressure, and mass flow rate data for water, hydrogen and oxygen are given in Table 12.3 according to their state numbers specified in Figure 12.5. The specific physical exergy and energy rates are calculated for each state, as presented in Table 12.3. The reference state values are taken to be 25 °C at the pressure of 100 kPa. The chemical exergy is associated with the departure of the chemical composition of a system from that of the environment. For simplicity, the chemical exergy considered in the analysis is rather a standard chemical exergy, based on the standard values of the dead state temperature of 25 °C and pressure of 100 kPa.

One of the main factors affecting the hydrogen production cost is the temperature of the electrolyzer. Figure 12.6 illustrates thermal and electricity supply to the system. Energy-efficiency values for the overall system vary between 80% and 87%, while exergy-efficiency values for that range from 79 to 86%. Also, it is clear from this figure that energy and exergy efficiencies increase with the electrolysis temperatures ranging from 473 to 1173 K.

The exergy demand of the system is illustrated in Figure 12.7. The total exergy demand of the system is constant with an increase of temperature, while electrical demand decreases, due to the exergy inlet of steam. The decrease

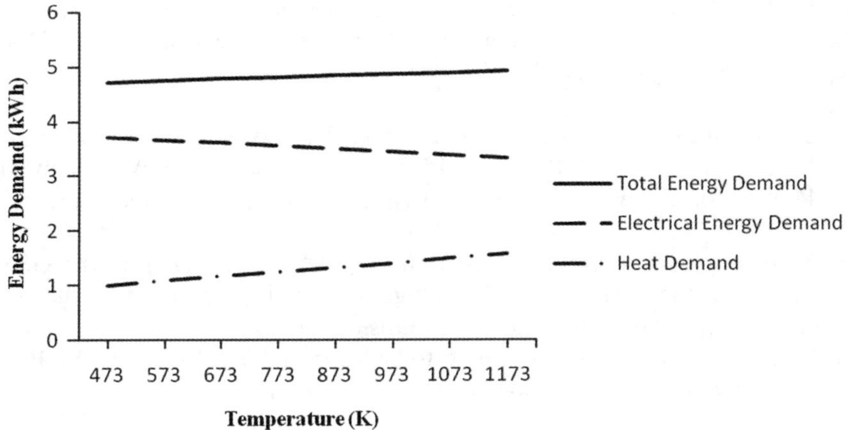

Figure 12.6 Thermal and electricity supply to the system in the case study 2.[1]

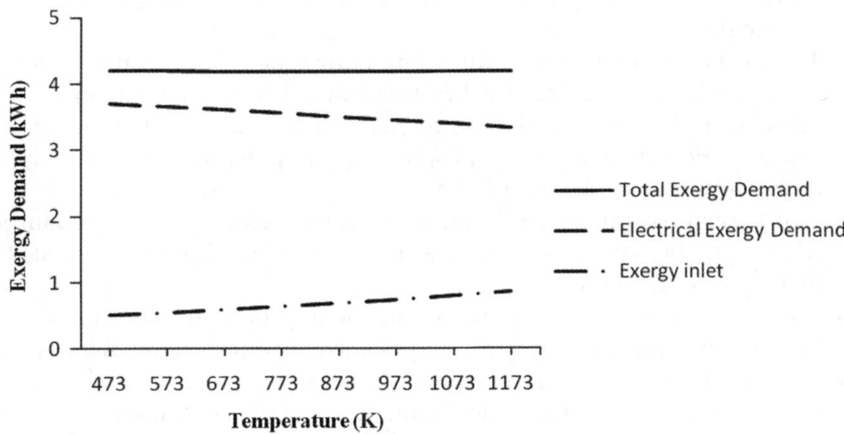

Figure 12.7 Exergy demand of the system in the case study 2.[1]

of electrical demand increases the thermal to hydrogen exergy conversion efficiency.

12.4 Conclusions

In this section, the relations for energetic and exergetic modeling of two compound energy systems are given for the system analysis and performance assessment. The two systems include a solar-assisted DHW tank integrated GSHP system and a HTE process coupled with a geothermal source. The performances of both systems are evaluated using energy and exergy analysis methods based on the experimental data. Energy and exergy specifications are presented in tables. The results obtained are presented in terms of energetic and exergetic aspects.

Some concluding remarks drawn from the results of the solar-assisted DHW integrated GSHP system are listed as follows:

- Exergy-efficiency values on a product/fuel basis were found to be 72.33% for the GSHP unit, 14.53% for the solar DHW system and 44.06% for the whole system at a dead (reference) state temperature of 19 °C.
- According to Van Gool's improvement potential rate ($\dot{IP}$) on the GSHP unit basis, the condenser had the highest $\dot{IP}$ value, followed by the compressor, the evaporator and the expansion valve.
- Exergetic COP values were found to be 0.245 and 0.201 for the GSHP unit and the whole system, respectively.

As far as a HTE process coupled with a geothermal source is concerned, the following concluding remarks may be extracted:[1]

- Using electrolysis of water splitting still needs to be improved. With the present electrolysis process, it is expensive due to consuming more electricity.
- The current technological status of all pathways of hydrogen production from geothermal requires further research and development on how to implement these first and how to make them more cost effective and efficient. Exergy analysis is proposed as a potential tool to help in answering some of these questions.
- Geothermal steam-assisted high-temperature electrolysis in countries where it is abundant due to geothermal sources could possibly reduce the hydrogen production cost.
- Low-temperature cycles can be coupled with geothermal sources, should be developed and analyzed in terms of their thermodynamic feasibilities.
- It is confirmed from the exergetic point of view that the electrolyzer temperature is one of the basic parameters that affects efficiency. Increase in electrolyzer temperature leads to an increase in exergy efficiency and a decrease in exergy destruction rate.

As a general conclusion, the following may be listed:

- There are various ways to describe exergy efficiency in the literature. In this regard, the use of the efficiency definition on the benefit/fuel basis is more convenient than that on the output/input basis.
- Exergy analysis is a useful tool for determining the locations, types and true magnitudes of energy losses, and therefore helps in the design of more efficient energy systems. It is also a way to a sustainable development and reveals whether or not (and by how much) it is possible to improve compound energy systems by reducing their inefficiencies.
- For further work, it is recommended to perform an exergoeconomic analysis, which is a combination of exergy and economics, and provides useful insights into the relations between thermodynamics and economics.

Furthermore, the magnitude of every significant thermodynamic inefficiency (exergy destruction and exergy loss) in the system has to be justified by considerations related to costs. In this regard, a thermoeconomic (exergoeconomic) analysis, which is a combination of thermodynamic and economic analyses, may aim at achieving this goal.

Acknowledgments

The author gratefully acknowledges the support provided by Ege University, Aksaray University, University of Ontario Institute of Technology (UOIT) and the Natural Sciences and Engineering Research Council of Canada. He also would like to express his appreciation to his wife (Fevziye Hepbasli) and his daughter (Nesrin Hepbasli) for their continued patience, understanding and full support throughout the preparation of this part as well as all the others. The continuous incentive support given by the Scientific & Technological Research Council of Turkey (TUBITAK) is also gratefully acknowledged, while the author is grateful to Prof. Dr. Ibrahim Dincer, presently working at UOIT in Canada, due to his continued support.

Nomenclature

A: surface area m^2
C: specific heat kJ/kg K
COP: heating coefficient of performance
$\dot{E}$: energy rate kW
ex: specific exergy kJ/kg
E_x: exergy amount kJ
$\dot{E}_x$: exergy rate kW
$\dot{F}$: exergy rate of fuel kW; factor
f: exergetic factor
G: Gibbs function kJ

h: specific enthalpy kJ/kg
$\bar{g}_f^0$: specific Gibbs free energy of formation kJ/kmol
$\bar{h}$: specific enthalpy kJ/kmol
$\bar{h}^0$: specific enthalpy at reference state kJ/kmol
$\bar{h}_f^0$: specific enthalpy of formation kJ/kmol
HHV: higher heating (gross calorific) value kJ/kg
I: current (A), global irradiance W/m^2
$\dot{I}$: rate of irreversibility, rate of exergy consumption kW
$I\dot{P}$: rate of improvement potential kW
$\dot{M}$: mass flow rate kg/s
M: molar mass kg/kmol
N: number of moles kmol
P: pressure kPa
$\dot{P}$: exergetic product rate kW
Q: heat-transfer amount kJ
$\dot{Q}$: heat-transfer rate kW
r: recycling ratio
R: gas constant kJ/kg K
s: specific entropy kJ/kg K
$\dot{V}$: entropy rate kW/K
T: temperature °C or K
V: voltage V
W: work kJ
$\dot{W}$: work rate (or power) kW

Greek letters

ω: specific humidity ratio (kg water/kg air)
η: energy (first law) efficiency
ψ: flow (specific) exergy kJ/kg
Δ: interval
δ: fuel-depletion rate
ε: exergy (second law) efficiency
ξ: productivity lack
χ: relative irreversibility
$\cos \varphi$: power factor

Subscripts

0: reference or dead state
A: air
act: actual
$Comp$: compressor
$Cond$: condenser
$dest$: destroyed
$dhwt$: domestic hot water tank

Elec:	electrical
Evap:	evaporator
Ex:	exergetic
Exp:	expansion valve
f:	free
Fh:	floor heating
fhs:	floor-heating system
G:	geothermal
Gen:	generation
Ghe:	ground heat exchanger
GSHP:	ground source heat pump
HE:	heat exchanger
In:	input, inlet
int:	internal
Mech:	mechanical
out:	output, outlet
P:	pressure
P:	product
per:	perfect
prt:	pressure relief tank
R:	reactant
r:	refrigerant
s:	isentropic process
scol:	solar collector
sr:	solar radiation
sys:	system
T:	total
U:	useful
V:	vapor
W:	water
1:	initial state
2:	final state

Superscripts

over dot:	quantity per unit time
over bar:	quantity per unit mole
0:	standard reference states
Ch:	chemical
pt:	potential
Kn:	kinetic
Ph:	physical

Abbreviations

DHW:	domestic hot water

GSHP: ground source heat pump
HT: high temperature
HTE: high-temperature electrolysis
HTSE: high-temperature steam electrolysis
LT: low temperature
MT: medium temperature

References

1. M. T. Balta, I. Dincer and A. Hepbasli, Thermodynamic assessment of geothermal energy use in hydrogen production, *Int. J. Hydrogen Energy*, 2009, **34**(7), 2922–2936.
2. H. Doukas, A. G. Papadopoulou, C. Nychtis, J. Psarras and N. van Beeck, Energy research and technology development data collection strategies: The case of Greece, *Renew. Sustain. Energy Rev.*, 2009, **13**, 682–688.
3. European Commission, Action plan for energy efficiency: Realizing the potential, COM(2006)545 final, 19 October 2006.
4. A. B. Omer, Energy, environment and sustainable development, *Renew. Sustain. Energy Rev.*, 2008, **12**, 2265–3000.
5. I. Dincer, The role of exergy in energy policy making, *Energy Policy*, 2002, **30**, 137–149.
6. K. Alqnne and A. Saari, Distributed energy generation and sustainable development, *Renew. Sustain. Energy Rev.*, 2006, **10**(6), 539–558.
7. A. Hepbasli, A key review on exergetic analysis and assessment of renewable energy resources for a sustainable future, *Renew. Sustain. Energy Rev.*, 2008, **12**(3), 593–661.
8. A. Hepbasli, Exergetic modeling and assessment of solar assisted domestic hot water tank integrated ground-source heat pump systems for residences, *Energy Buildings*, 2007, **39**(12), 1211–1217.
9. M. A. Rosen, M. N. Le and I. Dincer, Efficiency analysis of a cogeneration and district energy system, *Appl. Thermal Eng.*, 2005, **25**(1), 147–159.
10. I. Dincer, M. M. Hussain and I. Al-Zaharnah, Energy and exergy use in public and private sector of Saudi Arabia, *Energy Policy*, 2004, **32**(141), 1615–1624.
11. M. A. Rosen and I. Dincer, Exergy methods for assessing and comparing thermal storage systems, *Int. J. Energy Res.*, 2003, **27**(4), 415–430.
12. M. J. Moran, *Availability Analysis: A Guide to Efficiency Energy Use,* Englewood Cliffs, Prentice-Hall, NJ, 1982.
13. I. Dincer, M. M. Hussain and I. Al-Zaharnah, Energy and exergy use in public and private sector of Saudi Arabia, *Energy Policy*, 2004, **32**(141), 1615–1624.
14. F. Balkan, N. Colak and A. Hepbasli, Performance evaluation of a triple effect evaporator with forward feed using exergy analysis, *Int. J. Energy Res.*, 2005, **29**, 455–470.

15. G. Wall, Exergy tools, *Proc. Inst. Mech. Eng.*, 2003, **125**, 36.
16. A. Hepbasli and O. Akdemir, Energy and exergy analysis of a ground source (geothermal) heat pump system, *Energy Conver. Manag.*, 2004, **45**, 737–753.
17. O. Ozgener and A. Hepbasli, Experimental performance analysis of a solar assisted ground-source heat pump greenhouse heating system, *Energy Buildings*, 2005, **37**, 101–110.
18. T. J. Kotas, *The Exergy Method of Thermal Power Plants,* Krieger Publishing Company, Malabar, Florida, 1995.
19. Y. Cengel and M. A. Boles, *Thermodynamics: An Engineering Approach, 4th edn,* McGraw-Hill, New York, 2001.
20. J. Szargut, *Exergy Method: Technical and Ecological Applications,* WIT Press, Southampton Boston, 2005.
21. W. J. Wepfer, R. A. Gaggioli and E. F. Obert, Proper evaluation of available energy for HVAC, *ASHRAE Trans.*, 1979, **85**(1), 214–230.
22. R. DiPippo, Second law assessment of binary plants generating power from low-temperature geothermal fluids, *Geothermics*, 2004, **33**, 565–586.
23. W. Van Gool, *Energy policy fairly tales and factualities*, In: *Innovation and Technology-Strategies and Policies*, ed. O.D.D. Soares., A. Martins da Cruz, G. Costa Pereira, I.M.R.T. Soares, A.J.P.S. Reis, Kluwer, Dordrecht, 1997, pp. 93–105.
24. J. Y. Xiang, M. Cali and M. Santarelli, Calculation for physical and chemical exergy of flows in systems elaborating mixed-phase flows and a case study in an IRSOFC plant, *Int. J. Energy Res.*, 2004, **28**, 101–115.
25. V. Trillat-Berdal, B. Souyri and G. Fraisse, Experimental study of a ground-coupled heat pump combined with thermal solar collectors, *Energy Buildings*, 2006, **38**(12), 1477–1484.
26. R. Petela, Exergy of undiluted thermal radiation, *Sol. Energy*, 2003, **74**, 469–488.
27. A. Bejan, *Entropy Generation through Heat and Fluid Flow,* John Wiley & Sons, New York, 1994.
28. J. Sigurvinsson, C. Mansilla, B. Arnason, A. Bontemps, A. Maréchal, T. I. Sigfusson and F. Werkoff, Heat transfer problems for the production of hydrogen from geothermal energy, *Energy Conver. Manag.*, 2006, **47**, 3543–3451.
29. J. Sigurvinsson, C. Mansilla, P. Lovera and F. Werkoff, Can high temperature steam electrolysis function with geothermal heat?, *Int. J. Hydrogen Energy*, 2007, **32**, 1174–1182.
30. National Institute of Standards and Technology (NIST), http://webbook.nist.gov/chemistry/; Access date: 16 December 2009.
31. The Exergoecology Portal, http://www.exergoecology.com/excalc/; Access date: 16 December 2009.

Subject Index